STUDY GUIDE

LINEAR ALGEBRA and Its Applications

David C. Lay
University of Maryland

Addison-Wesley Publishing Company
Reading, Massachusetts • Menlo Park, California • New York
Don Mills, Ontario • Wokingham, England • Amsterdam • Bonn
Sydney • Singapore • Tokyo • Madrid • San Juan • Milan • Paris

MATLAB is a registered trademark of The MathWorks, Inc.

Reproduced by Addison-Wesley from camera-ready copy supplied by the author.

ISBN 0-201-52033-8

1 2 3 4 5 6 7 8 9 10 VG 97969594

TABLE OF CONTENTS

HOW TO STUDY LINEAR ALGEBRA

This Study Guide is designed to help you improve your performance in your linear algebra course. Students who have used preliminary versions of this material have told me how much it helped them study and prepare for tests. The first time I used these notes, I began distributing them about halfway through the semester. Grades on the next exam were substantially higher. For some students, the improvement was dramatic.

Main Features

Key Ideas and **Study Notes** highlight important facts and guide you through each section, with suggestions for learning the material and relating it to earlier ideas. In some sections, summaries will help you organize the material for exams.

Detailed solutions of every third odd exercise are given, plus complete explanations for any odd-numbered exercise whose answer in the text is only a "Hint". I designed the exercise sets with the Study Guide in mind, making sure that exercises numbered 1, 7, 13, 19, ... covered every major concept. However, please heed the warning at the end of this introduction.

Study Tips point out particularly important exercises, give hints on what and how to study, and sometimes identify potential exam questions.

Frequent Warnings describe common errors and show you how to prevent them.

Checkpoints (questions) are placed at strategic points to help you check how well you are understanding and retaining the material. The answer for a checkpoint is at the end of the section in which it appears.

MATLAB boxes discuss the MATLAB commands as they are needed for the exercises. I encourage you to try MATLAB, because it is so easy to learn. If MATLAB is already installed on computers at your school, about all you need to know is how to start and stop the MATLAB program. To save time on homework, send for the MATLAB Toolbox that has the data for over 700 exercises.

Steps to Success in Linear Alagebra

1. **Read each section thoroughly before you begin the exercises.** Most students are not used to doing this in courses that precede linear algebra. They could survive by looking at the examples only when they were unable to work an exercise. That simply will not work in linear algebra. If you "copy" an example (with necessary modifications), you may think you are "understanding" the problem, but very little true learning will take place. (You'll find that out on your first exam.) After reading the

text, use the "Key Ideas" and "Study Notes" in this *Guide* to help you work through the text again. *Then* start on the exercises! In the long run, this strategy will improve your performance and save you time.

2. **Prepare for each class as you would for a language class.** You must work on linear algebra between every class meeting to avoid getting behind. *This is extremely important!* Most sections in the text build on preceding sections, and once you are behind, catching up with the class is often difficult. You'll probably encounter one or two concepts (and several new terms) in each class that must be studied in order for the discussion the next day to make sense. The fact that concepts may seem "simple" does not mean you can afford to postpone your study until the weekend. The homework may be harder than you expect. The most valuable advice I can give you is to keep up with the course.

3. **Concentrate more on learning definitions, facts, and concepts, than on practicing routine computations or algorithms.** Seek connections *between* concepts. Many theorems and boxed "facts" describe such connections. For instance, see Theorem 2 in Section 1.2, and Theorems 1 and 2 in Section 2.2. Your goal is to think in general terms, to focus on the principles behind the specific computations. You will have to read each section slowly several times. Practice writing explanations for the homework exercises, using complete sentences. Don't just talk to yourself and *say* what you would write if asked on an exam.

4. **Review frequently.** Use the Glossary Checklist at the end of the chapter to study the important terms. Learning definitions is the first step towards a full understanding of basic concepts. For each main concept, collect related facts and write them on one or more sheets of paper. The writing will help you to learn because it encourages you to look at material slowly and to think about it while you write. Read through earlier sections that may have related material, such as key examples or exercises. Summarize them on your paper. You should find the Study Guide really helpful in your review.

WARNING Because you can find complete solutions here to many problems, you will be tempted to read the explanations before you really try to write out the solutions yourself. Don't do it! If you merely think a bit about a problem and then check to see if your idea is basically correct, you are likely to overestimate your understanding. Some of my students have done this and miserably failed the first exam. By then the damage was done, and they had great difficulty catching up with the class. Proper use of the Study Guide, however, will help you to succeed and enjoy the course at the same time.

1

Systems of Linear Equations

1.1 INTRODUCTION TO SYSTEMS OF LINEAR EQUATIONS

The fundamental concepts presented in this section and the next must be mastered, for they will be used throughout the course.

STUDY NOTES

Please read **How to Study Linear Algebra**, pages v-vi, before you continue.

The text uses boldface type to identify important terms the first time they appear. You need to learn them; some students write selected terms on 3×5 cards, for review. At the end of each chapter in this Study Guide, you will find a glossary checklist that gives the main terms and helps you learn their definitions.

The text defines the **size** of a matrix. Don't use the term *dimension*, even though that appears in some computer programming languages, because in linear algebra, *dimension* refers to another concept (in Section 5.5).

The first few examples are so simple that they could be solved by a variety of techniques. But it is important to learn the systematic method presented here, because it easily handles more complicated linear systems, and it works in all cases. Read the solution of Example 1 at least twice: first study the equations, and then study the augmented matrices.

The calculations in this section are based on the following important fact. When elementary row operations are applied to a linear system, the new system has exactly the same solution set, because each operation is reversible. (See the text.) The steps in the summary below will be modified slightly in Section 1.2.

Summary of the Elimination Method (for This Section)

1. The first equation must contain an x_1. Interchange equations, if necessary. This will create a nonzero entry in the first row, first column, of the augmented matrix.

2. Eliminate x_1 terms in the other equations. That is, use replacement operations to create zeros in the first column of the matrix below the first row.
3. Obtain an x_2 term in the second equation. (Interchange the second equation with one below, if needed, but don't touch the first equation.) You may scale the second equation, if desired, to create a 1 in the second column and second row of the matrix.
4. Eliminate x_2 terms in equations below the second equation, using replacement operations.
5. Continue with x_3 in the third equation, x_4 in the fourth equation, etc., eliminating these variables in the equations below. This will produce a "triangular" system (at least for systems in this section).
6. Check if the system in triangular form is consistent. If it is, a solution is found by starting with the last nonzero equation and working back up to the first equation. Each variable on the "diagonal" is used to eliminate the terms in that variable above it. The solution to the system becomes apparent when the system is finally transformed into "diagonal" form.
7. Check any solutions you find by substituting them into the original system.

The *solution set* of a system of linear equations either is empty, or contains one solution, or contains infinitely many solutions. When asked to "solve" a system, you may write "inconsistent" if the system has no solution.

As you will see later, determining the "size" of the solution set is sometimes more important than actually computing the solution or solutions. For that reason, pay close attention to the subsection on existence and uniqueness questions. Be sure to work at least two of Exercises 27—30.

SOLUTIONS TO EXERCISES

Get into the habit *now* of studying the Practice Problems before you start the exercises. To prepare for the exercises in this section, you should write the augmented matrices for the systems in Practice Problems 1 and 2, and try to determine the next row operations, without reading the Practice Problem solutions in the text.

1. $\begin{aligned} x_1 + 7x_2 &= 4 \\ -2x_1 - 9x_2 &= 2 \end{aligned}$ $\begin{bmatrix} 1 & 7 & 4 \\ -2 & -9 & 2 \end{bmatrix}$

Replace row 2 by row 2 plus 2·row 1: $\begin{aligned} x_1 + 7x_2 &= 4 \\ 5x_2 &= 10 \end{aligned}$ $\begin{bmatrix} 1 & 7 & 4 \\ 0 & 5 & 10 \end{bmatrix}$

Scale row 2, multiply by 1/5: $\begin{aligned} x_1 + 7x_2 &= 4 \\ x_2 &= 2 \end{aligned}$ $\begin{bmatrix} 1 & 7 & 4 \\ 0 & 1 & 2 \end{bmatrix}$

Replace row 1 by row 1 + (-7)·row 2: $\begin{aligned} x_1 \quad &= -10 \\ x_2 &= 2 \end{aligned}$ $\begin{bmatrix} 1 & 0 & -10 \\ 0 & 1 & 2 \end{bmatrix}$

The solution is (-10,2). Check:

$$\begin{aligned} (-10) + 7(2) &= -10 + 14 = 4 \\ -2(-10) - 9(2) &= 20 - 18 = 2 \end{aligned}$$

7. $\begin{bmatrix} 1 & 3 & 0 & 5 & -5 \\ 0 & 1 & -6 & 9 & 0 \\ 0 & 0 & 2 & 7 & 1 \\ 0 & 0 & 1 & 4 & -2 \end{bmatrix}$. To simplify hand computation, the next step is either to interchange rows 3 and 4 or to multiply row 3 by 1/2. Another possibility is to replace row 4 by row 4 + (-1/2)·row 3.

13. $\begin{bmatrix} 1 & -1 & 0 & 0 & -5 \\ 0 & 1 & -2 & 0 & -7 \\ 0 & 0 & 1 & -3 & -2 \\ 0 & 0 & 0 & 1 & 4 \end{bmatrix}$

Replace row 3 by row 3 + 3·row 4: $\begin{bmatrix} 1 & -1 & 0 & 0 & -5 \\ 0 & 1 & -2 & 0 & -7 \\ 0 & 0 & 1 & 0 & 10 \\ 0 & 0 & 0 & 1 & 4 \end{bmatrix}$

Replace row 2 by row 2 + 2·row 3: $\begin{bmatrix} 1 & -1 & 0 & 0 & -5 \\ 0 & 1 & 0 & 0 & 13 \\ 0 & 0 & 1 & 0 & 10 \\ 0 & 0 & 0 & 1 & 4 \end{bmatrix}$

Replace row 1 by row 1 + row 2: $\begin{bmatrix} 1 & 0 & 0 & 0 & 8 \\ 0 & 1 & 0 & 0 & 13 \\ 0 & 0 & 1 & 0 & 10 \\ 0 & 0 & 0 & 1 & 4 \end{bmatrix}$

The solution is (8,13,10,4), a list of four numbers. Note that the solution set contains only one *solution*, not four.

19. Work with the augmented matrix: $\begin{bmatrix} 0 & 1 & 5 & -4 \\ 1 & 4 & 3 & -2 \\ 2 & 7 & 1 & -1 \end{bmatrix}$.

Interchange rows 1 and 2: $\begin{bmatrix} 1 & 4 & 3 & -2 \\ 0 & 1 & 5 & -4 \\ 2 & 7 & 1 & -1 \end{bmatrix}$

Add (-2)·row 1 to row 3: $\begin{bmatrix} 1 & 4 & 3 & -2 \\ 0 & 1 & 5 & -4 \\ 0 & -1 & -5 & 3 \end{bmatrix}$

Add row 2 to row 3: $\begin{bmatrix} 1 & 4 & 3 & -2 \\ 0 & 1 & 5 & -4 \\ 0 & 0 & 0 & -1 \end{bmatrix}$ $\quad \begin{aligned} x_1 + 4x_2 + 3x_3 &= -2 \\ x_2 + 5x_3 &= -4 \\ 0 &= -1 \end{aligned}$

The equation 0 = -1 is never true, so the solution set of this triangular system is empty. Since the original system has the same solution set, the original system is inconsistent (has no solution).

Study Tip: When writing a coefficient matrix or augmented matrix for a system of linear equations, be sure that the variables appear in all equations *in the same order*. Arrange the variables in columns, as in the text; place zeros in the matrix whenever a variable is missing from an equation.

25. Work with the augmented matrix: $\begin{bmatrix} 0 & 2 & 2 & 0 & 0 \\ 1 & 0 & 0 & -2 & -3 \\ 0 & 0 & 1 & 3 & -4 \\ -2 & 3 & 2 & 1 & 5 \end{bmatrix}$.

Interchange rows 1 and 2: $\begin{bmatrix} 1 & 0 & 0 & -2 & -3 \\ 0 & 2 & 2 & 0 & 0 \\ 0 & 0 & 1 & 3 & -4 \\ -2 & 3 & 2 & 1 & 5 \end{bmatrix}$

Add 2·row 1 to row 4: $\begin{bmatrix} 1 & 0 & 0 & -2 & -3 \\ 0 & 2 & 2 & 0 & 0 \\ 0 & 0 & 1 & 3 & -4 \\ 0 & 3 & 2 & -3 & -1 \end{bmatrix}$

Add (-3/2)·row 2 to row 4: $\begin{bmatrix} 1 & 0 & 0 & -2 & -3 \\ 0 & 2 & 2 & 0 & 0 \\ 0 & 0 & 1 & 3 & -4 \\ 0 & 0 & -1 & -3 & -1 \end{bmatrix}$

Add 1·row 3 to row 4:
$$\begin{bmatrix} 1 & 0 & 0 & -2 & -3 \\ 0 & 2 & 2 & 0 & 0 \\ 0 & 0 & 1 & 3 & -4 \\ 0 & 0 & 0 & 0 & -5 \end{bmatrix}$$

The system is inconsistent, because $0 \neq -5$.

Study Tip: Pay attention to how a problem is worded. If you are instructed only to determine existence or uniqueness of a solution, as in Exercise 25, stop row operations when you reach a "triangular" form.

31. $\begin{bmatrix} 1 & 3 & -1 \\ 0 & 2 & -4 \\ 0 & -3 & 4 \end{bmatrix}, \begin{bmatrix} 1 & 3 & -1 \\ 0 & 1 & -2 \\ 0 & -3 & 4 \end{bmatrix}$. Multiply row 2 by 1/2; multiply row 2 by 2.

MATLAB To obtain the data for the exercises in this section while you are running MATLAB, type **x1s1** , and press <Enter>. You will then see a list of exercises for which data are available. Type the number of the appropriate exercise (and press <Enter>).

In this exercise set, the data for each exercise are stored in a matrix *M*. Row operations on *M* are performed by the following commands (on the Toolbox data diskette for this text, available from Mathworks):

replace(M,r,m,s)	Replaces row r of M by row r + m·row s
swap(M,r,s)	Interchanges rows r and s of M
scale(M,r,c)	Multiplies row r of M by a nonzero scalar c

(Press <Enter> after each MATLAB command, displayed in boldface type.) The name of any matrix in your MATLAB workspace can be inserted in place of *M*; the letters *r*, *m*, *s*, and *c* stand for numbers you choose.

If you enter one of these commands, say, **swap(M,1,3)**, then the new matrix, produced from *M*, is stored in the matrix "ans" (for "answer"). If, instead, you type **M1 = swap(M,1,3)**, then the answer is stored in a new matrix *M*1. If the next operation is **M2 = replace(M1,2,5,1)**, then the result of changing *M*1 is placed in *M*2, and so on.

The advantage of giving a new name to each new matrix is that you can easily go back a step if you don't like what you just did to a matrix. If, instead, you type **M = replace(M,2,5,1)** , then the result is placed back in *M* and the "old" *M* is lost. Of course, the "reverse" operation, **M = replace(M,2,-5,1)** will bring back the old *M*.

Note: For the simple problems in this section and the next, the multiple *m* you need in the command **replace(M,r,m,s)** will usually be a small integer or fraction that you can compute in your head. In general, *m* may not be so easy to compute mentally. In Section 2.3, I'll describe a simple way to write *m* in terms of the entries in *M*.

Warning: Using MATLAB is fun and will save you time, but make sure you can perform row operations rapidly and accurately with pencil and paper. Probably, you should work all the exercises in Section 1.1 by hand and only use MATLAB to check your work.

1.2 ROW REDUCTION AND ECHELON FORMS

Our interest in the row reduction algorithm lies mostly in the echelon forms that are created by the algorithm. For practical work, a computer should perform the calculations. However, you need to understand the algorithm so you can learn how to use it for various tasks. Also, unless you take your exams at a computer, you must be able to perform row reduction quickly and accurately by hand.

STUDY NOTES

The row reduction algorithm applies to any matrix, not just an augmented matrix for a linear system. In many cases, all you need is an echelon form. The reduced echelon form is mainly used when it comes from an augmented matrix and you have to find all the solutions of a linear system.

Strategies for faster and more accurate row reduction:

- Avoid subtraction in a row replacement. It leads to mistakes in arithmetic. Instead, add a negative multiple of one row to another.
- Always enclose each matrix with brackets or large parentheses.
- To save time, combine all row replacement operations that use the *same* pivot position, and write just one new matrix. Never "clean out" more than one column at a time. (You can combine several scaling operations, or combine several interchanges, if you are careful. But that seldom will be necessary.)
- *Never* combine an interchange with a replacement. In general, don't combine different types of row operations. This will be particularly important when you evaluate determinants, in Chapters 4 and 6.

How to avoid copying errors:

- Practice neat writing, not too small. Develop proper habits in homework so your work on tests will be accurate and complete.
- Write a matrix row by row. If possible, place the new matrix beside the old one. Your eye is less likely to read from the wrong row if

you place the new matrix beside the old one. Arrange your sequence of matrices across the page, rather than down the page.

- Don't let your work flow from one side of a paper to the reverse side.

Study Tips: Theorem 2 is a key result for future work. Read it carefully. Also, study the procedure in the box following Theorem 2. Failure to write out the system of equations (step 4) is a common source of errors.

Practice Problem 1 identifies a common student error. One way to avoid the mistake discussed there is to assign a new variable, say s, as the parameter, and write

$$\begin{cases} x_1 = 9 + 2s \\ x_2 = 3 - s \\ x_3 = s, \quad \text{where } s \text{ is a parameter.} \end{cases}$$

You may do this if you wish, but it takes more effort.

SOLUTIONS TO EXERCISES

1. To check whether a matrix is in echelon form ask the questions:

(i) *Is every nonzero row above the all-zero rows (if any)?*

The matrix (d) fails this test, so it is *not* in echelon form. Do not apply the other tests to this matrix.

(ii) *Are the leading entries in a stair-step pattern, with zeroes below each leading entry?*

The matrices (a), (b), (c) all pass tests (i) and (ii), so they are in echelon form.

To check whether a matrix in echelon form is actually in *reduced* echelon form, ask two more questions:

(iii) *Is there a 1 in every pivot position?*

The matrices (a), (b), (c) all pass this test, so ask:

(iv) *Is each leading 1 the only nonzero entry in its column?*

Matrices (a) and (b) pass all four tests, so they are in reduced echelon; but matrix (c) is not in reduced echelon form.

7. $\begin{bmatrix} 1 & 0 & 2 & 5 \\ 2 & 0 & 3 & 6 \end{bmatrix} \sim \begin{bmatrix} 1 & 0 & 2 & 5 \\ 0 & 0 & -1 & -4 \end{bmatrix} \sim \begin{bmatrix} 1 & 0 & 2 & 5 \\ 0 & 0 & 1 & 4 \end{bmatrix} \sim \begin{bmatrix} 1 & 0 & 0 & -3 \\ 0 & 0 & 1 & 4 \end{bmatrix}$

The corresponding system of equations is: $\begin{aligned} x_1 \quad\quad &= -3 \\ x_3 &= 4 \end{aligned}$.

The basic variables are x_1 and x_3, corresponding to the pivot columns 1 and 3 in the matrix. Any other variables are free, so the general solution is $\begin{cases} x_1 = -3 \\ x_2 \text{ is free.} \\ x_3 = 4 \end{cases}$ Two incorrect answers are $\begin{cases} x_1 = -3 \\ x_2 = 0 \\ x_3 = 4 \end{cases}$ and $\begin{cases} x_1 = -3 \\ x_3 = 4 \end{cases}$ (with nothing said about x_2).

Warning: The remarks for Exercise 7 are important, because students often have difficulty with a problem in which a variable does not appear in the equation.

13. Combine two replacement operations in the first step:

$$\begin{bmatrix} 2 & -4 & 3 \\ -6 & 12 & -9 \\ 4 & -8 & 6 \end{bmatrix} \sim \begin{bmatrix} 2 & -4 & 3 \\ 0 & 0 & 0 \\ 0 & 0 & 0 \end{bmatrix} \sim \begin{bmatrix} 1 & -2 & 3/2 \\ 0 & 0 & 0 \\ 0 & 0 & 0 \end{bmatrix} \qquad \begin{aligned} x_1 - 2x_2 &= 3/2 \\ 0 &= 0 \\ 0 &= 0 \end{aligned}$$

Thus $x_1 = 3/2 + 2x_2$, and the general solution is $\begin{cases} x_1 = 3/2 + 2x_2 \\ x_2 \text{ is free} \end{cases}$.

19. The matrix given is not yet in reduced echelon form. So compute

$$\begin{bmatrix} 1 & -2 & 0 & 0 & 7 & -3 \\ 0 & 1 & 0 & 0 & -3 & 1 \\ 0 & 0 & 0 & 1 & 5 & -4 \\ 0 & 0 & 0 & 0 & 0 & 0 \end{bmatrix} \sim \begin{bmatrix} 1 & 0 & 0 & 0 & 1 & -1 \\ 0 & 1 & 0 & 0 & -3 & 1 \\ 0 & 0 & 0 & 1 & 5 & -4 \\ 0 & 0 & 0 & 0 & 0 & 0 \end{bmatrix} \qquad \begin{aligned} x_1 \qquad\qquad + x_5 &= -1 \\ x_2 \qquad - 3x_5 &= 1 \\ x_4 + 5x_5 &= -4 \end{aligned}$$

Columns 1,2, and 4 are pivot columns, and x_1, x_2, and x_4 are the basic variables. The other variables, x_3 and x_5, are free. Solve for the basic variables, and write the general solution as

$$\begin{cases} x_1 = -1 - x_5 \\ x_2 = 1 + 3x_5 \\ x_3 \text{ is free} \\ x_4 = -4 - 5x_5 \\ x_5 \text{ is free} \end{cases}$$

25. $\begin{bmatrix} 1 & h & 3 \\ 2 & 8 & 1 \end{bmatrix} \sim \begin{bmatrix} 1 & h & 3 \\ 0 & 8-2h & -5 \end{bmatrix}$. There are only two possibilities:

(1) $8 - 2h = 0$. In this case $\begin{bmatrix} 1 & h & 3 \\ 0 & 0 & -5 \end{bmatrix}$ represents an inconsistent system. So the system has no solution when $2h = 8$, and $h = 4$.

(2) $8 - 2h \neq 0$. In this case the augmented matrix corresponds to a consistent system, and there are no free variables. So the system has a unique solution when $h \neq 4$.

Study Tip: Be sure to work on Exercises 22-27. The experience will help you later. These exercises make nice quiz questions, too.

31. The full solution is in the text.

33. If a linear system is consistent, then the solution is unique if and only if every column in the coefficient matrix is a pivot column.

This statement is true because the free variables correspond to *non-pivot* columns of the coefficient matrix. The columns are all pivot columns if and only if there are no free variables. And there are no free variables if and only if the solution is unique, by Theorem 2.

Study Tip: Notice from Exercise 33 that the question of uniqueness of the solution of a linear system is not influenced by the numbers in the right-most column of the augmented matrix.

Appendix: A Mathematical Note

You need to know what the phrase "if and only if" means. It was used above in Exercise 33, and it will appear often in Theorems and in boxed facts. The phrase "if and only if" always appears between two complete statements. Look at Theorem 2, for instance:

$$\left\{\begin{array}{c}\text{A specific}\\ \text{linear system}\\ \text{is consistent}\end{array}\right\} \quad \text{if and only if} \quad \left\{\begin{array}{l}\text{the rightmost column of the}\\ \text{augmented matrix of the linear}\\ \text{system is not a pivot column.}\end{array}\right\} \qquad (1)$$

The entire sentence means that the two statements in parentheses are either both true or both false.

Sentence (1) has the general form

$$P \textit{ if and only if } Q \qquad (2)$$

where P denotes the first statement and Q denotes the second statement. The sentence (2) says two things:

If statement P is true, then statement Q is true, too. (3)

If statement Q is true, then statement P is also true. (4)

The mathematical shorthand for (3) is "$P \Rightarrow Q$" and is read as "P implies Q." Similarly, (4) is written as "$Q \Rightarrow P$". The abbreviation for (2) is "$P \Leftrightarrow Q$". Another abbreviation is "P *iff* Q".

1.3 APPLICATIONS OF LINEAR SYSTEMS

This section is the first of eleven sections devoted to uses of linear algebra. In addition, brief discussions of applications appear in many other sections. The applications included in the text were selected to give you an impression of the power of linear algebra. You are likely to encounter some of these topics again—in school or in your career. The discussions in the text should help you whenever you need to use linear algebra techniques.

KEY IDEAS

Traffic Flow: Assign a variable to each branch of the network. The flow into a junction equals the flow out of the junction; so each junction produces one equation. An additional equation expresses the fact that the total flow into the network equals the total flow out of the network. Make sure the variables are aligned in columns when you rearrange the equations into a system of linear equations.

Electrical Networks: Assign a branch current variable and a current flow direction to each branch. At the end of the problem, if the current flow turns out to be negative, then the current flows in the direction opposite to the one you chose. Kirchhoff's voltage law produces one equation for each loop in the network:

$$\left\{\begin{matrix}\text{Sum of the } RI \text{ voltage drops}\\ \text{across resistors in the loop}\end{matrix}\right\} = \left\{\begin{matrix}\text{Sum of all voltage}\\ \text{sources in the loop}\end{matrix}\right\}$$

To set up the equation, choose a direction around the loop. If this direction passes through a resistor or voltage source in the same direction as the current flow in the diagram, count the *RI* voltage drop (for a resistor) or the voltage gain (for a battery) as positive; otherwise treat the voltage drop or gain as negative. Write such an equation for each "minimal" loop in the network

Kirchhoff's current law produces one equation for each node:

$$\left\{\begin{matrix}\text{Sum of the currents}\\ \text{into the node}\end{matrix}\right\} = \left\{\begin{matrix}\text{Sum of the currents}\\ \text{out of the node}\end{matrix}\right\}$$

The electrical networks in this section have only two nodes, so only one equation is needed. (The other equation would give exactly the same relation among the currents.) When more nodes are present, you will need more node equations. More complicated networks are often analyzed using graph theory, an important tool in linear algebra. See the references below.

The Leontief Exchange Model: A country's economy is divided into various sectors. The *price* of a sector's output is the total dollar value of that sector's output for one year. Each column in an *exchange table* for the economy corresponds to one sector and describes the percentage distribution of the sector's output. The percentages are written as decimals, and the entries in each column sum to 1.

The basic problem is to determine the price for each sector's output in a way that the income of each sector (that is, the price of its output) should exactly equal its expenses. Each row of the exchange table gives the data necessary to compute the expenses of one sector.

STUDY NOTES

If you try to verify the row reduction in Example 1, interchange rows 1 and 4 as the first step. This will simplify your work. Placing the equation "$x_1 \quad + x_5 = 600$" at the top is better, because adding multiples of this row to rows below will not create new nonzero entries in columns 2 to 4.

Further Reading

R. Brualdi and H. J. Ryser, *Combinatorial Matrix Theory*, New York: Cambridge University Press, 1991.

L. R. Ford, Jr., and D. R. Fulkerson, *Flows in Networks*, Princeton: Princeton University Press, 1962.

Ben Noble and James W. Daniel, *Applied Linear Algebra*, 3rd edition, Englewood Cliffs: Prentice-Hall, 1988, pages 54-59.

SOLUTIONS TO EXERCISES

1. The total flow of 80 into the network must equal the total flow out of the network. So $20 + x_4 = 80$, and $x_4 = 60$. Next, use the figure to write an equation for each junction:

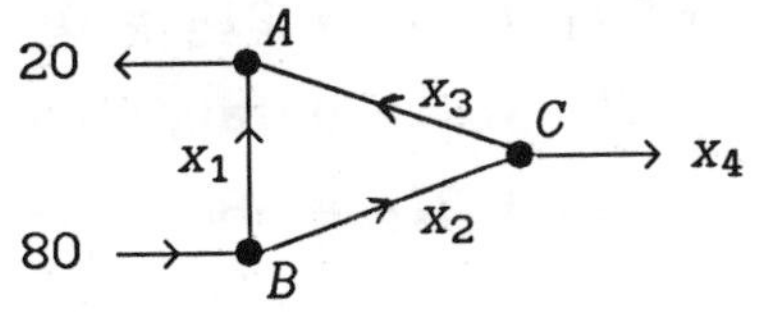

Junction	Flow in		Flow out
A	$x_1 + x_3$	=	20
B	80	=	$x_1 + x_2$
C	x_2	=	$x_3 + x_4$

Combine and arrange the equations:

$$\begin{aligned} x_1 \quad\quad + x_3 \quad\quad &= 20 \\ x_1 + x_2 \quad\quad\quad\quad &= 80 \\ x_2 - x_3 - x_4 &= 0 \\ x_4 &= 60 \end{aligned} \qquad \begin{bmatrix} 1 & 0 & 1 & 0 & 20 \\ 1 & 1 & 0 & 0 & 80 \\ 0 & 1 & -1 & -1 & 0 \\ 0 & 0 & 0 & 1 & 60 \end{bmatrix}$$

Row reduce the augmented matrix:

$$\sim \begin{bmatrix} 1 & 0 & 1 & 0 & 20 \\ 0 & 1 & -1 & 0 & 60 \\ 0 & 1 & -1 & -1 & 0 \\ 0 & 0 & 0 & 1 & 60 \end{bmatrix} \sim \begin{bmatrix} 1 & 0 & 1 & 0 & 20 \\ 0 & 1 & -1 & 0 & 60 \\ 0 & 0 & 0 & -1 & -60 \\ 0 & 0 & 0 & 1 & 60 \end{bmatrix} \sim \begin{bmatrix} 1 & 0 & 1 & 0 & 20 \\ 0 & 1 & -1 & 0 & 60 \\ 0 & 0 & 0 & 1 & 60 \\ 0 & 0 & 0 & 0 & 0 \end{bmatrix}$$

and solve for the basic variables in terms of the free variable x_3, to obtain the general flow pattern:

$$\begin{aligned} x_1 \quad + x_3 \quad &= 20 \\ x_2 - x_3 \quad &= 60 \\ x_4 &= 60 \end{aligned} \quad \Rightarrow \quad \begin{aligned} &x_1 = 20 - x_3 \\ &x_2 = 60 + x_3 \\ &x_3 \text{ is free} \\ &x_4 = 60 \end{aligned}$$

If all flows are nonnegative, then

$$\begin{aligned} &x_1 = 20 - x_3 \geq 0 \\ &x_2 = 60 + x_3 \geq 0 \\ &x_3 \geq 0 \end{aligned}$$

The first inequality shows that $20 \geq x_3$, so x_3 cannot exceed 20. The second inequality provides no further information about x_3.

7.

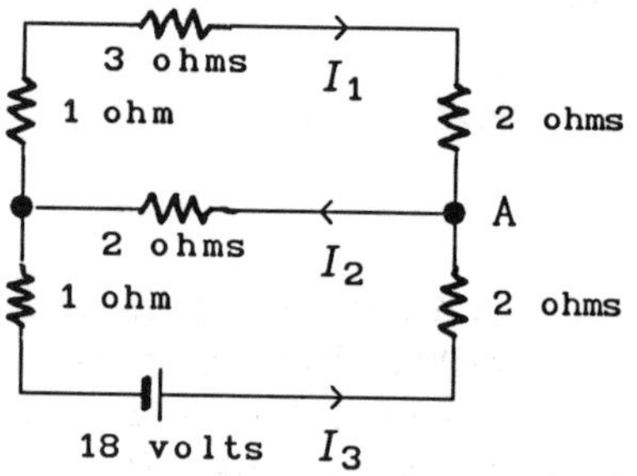

Kirchhoff's current law at A:

$$I_1 + I_3 = I_2$$

Kirchhoff's voltage law, uppper loop:

$$I_1 + 3I_1 + 2I_1 + 2I_2 = 0$$

Kirchhoff's voltage law, lower loop:

$$I_3 + 2I_3 + 2I_2 = 18$$

Rearrange the equations, and reduce the augmented matrix:

$$\begin{aligned} I_1 - I_2 + I_3 &= 0 \\ 6I_1 + 2I_2 \quad\quad &= 0 \\ 2I_2 + 3I_3 &= 18 \end{aligned} \qquad \begin{bmatrix} 1 & -1 & 1 & 0 \\ 6 & 2 & 0 & 0 \\ 0 & 2 & 3 & 18 \end{bmatrix} \sim \cdots \sim \begin{bmatrix} 1 & 0 & 0 & -1 \\ 0 & 1 & 0 & 3 \\ 0 & 0 & 1 & 4 \end{bmatrix}$$

The branch currents are $I_1 = -1$ amps (so the current flows in the direction opposite to that shown in the figure), $I_2 = 3$ amps, and $I_3 = 4$ amps.

13. a. Fill in the exchange table one column at a time. The entries in a column describe where a sector's output goes. The decimal fractions in each column sum to 1.

Distribution of Output From:				
Chemicals	Fuel	Machinery		Purchased by:
output ↓	↓	↓		
.2	.8	.4	input ⟶	Chemicals
.3	.1	.4	⟶	Fuels
.5	.1	.2	⟶	Machinery

b. Denote the total annual output (in dollars) of the sectors by p_c, p_f, and p_m. From the first row of the table, the total input to the Chemical & Metals sector is $.2p_c + .8p_f + .4p_m$. So the equilibrium prices must satisfy

$$\underset{\text{income}}{p_c} = \underset{\text{expenses}}{.2p_c + .8p_f + .4p_m}$$

From the second and third rows of the table, the income/expense requirements for the Fuels & Power sector and the Machinery sector are, respectively,

$$\begin{aligned} p_f &= .3p_c + .1p_f + .4p_m \\ p_m &= .5p_c + .1p_f + .2p_m \end{aligned}$$

Move all variables to the left side and combine like terms:

$$\begin{aligned} .8p_c - .8p_f - .4p_m &= 0 \\ -.3p_c + .9p_f - .4p_m &= 0 \\ -.5p_c - .1p_f + .8p_m &= 0 \end{aligned}$$

c. To obtain the reduced echelon form with MATLAB to solve this system, use the **Replace**, **Scale**, and **Swap** commands. Actually, hand calculations are not too messy. To simplify the calculations, first scale each row of the augmented matrix by 10, then continue as usual.

$$\begin{bmatrix} 8 & -8 & -4 & 0 \\ -3 & 9 & -4 & 0 \\ -5 & -1 & 8 & 0 \end{bmatrix} \sim \begin{bmatrix} 1 & -1 & -.5 & 0 \\ -3 & 9 & -4 & 0 \\ -5 & -1 & 8 & 0 \end{bmatrix} \sim \begin{bmatrix} 1 & -1 & -.5 & 0 \\ 0 & 6 & -5.5 & 0 \\ 0 & -6 & 5.5 & 0 \end{bmatrix} \sim$$

$$\begin{bmatrix} 1 & -1 & -.5 & 0 \\ 0 & 1 & -.917 & 0 \\ 0 & 0 & 0 & 0 \end{bmatrix} \sim \begin{bmatrix} 1 & 0 & -1.417 & 0 \\ 0 & 1 & -.917 & 0 \\ 0 & 0 & 0 & 0 \end{bmatrix}$$

The number of decimal places displayed is somewhat arbitrary.

The general solution is $p_c = 1.417p_m$, $p_f = .917p_m$, with p_m free. If p_m is assigned the value 100, then $p_c = 141.7$ and $p_f = 91.7$. Note that only the *ratios* of the prices are determined. This makes sense, for if the prices were converted from, say, dollars to marks or yen, the inputs and outputs of each sector would still balance. The economic equilibrium is not be affected by a proportional change in prices.

CHAPTER 1 GLOSSARY CHECKLIST

Check your knowledge by attempting to write definitions of the terms below. Then compare your work with the definitions given in the text's Glossary. Ask your instructor which definitions, if any, might appear on a test.

augmented matrix: A matrix made up of a ....

back-substitution (with matrix notation): The ... phase of row reduction of an ....

basic variable: A variable in a linear system that ....

coefficient matrix: A matrix whose entries are ....

consistent linear system: A linear system with ....

echelon form (or **row echelon form**, of a matrix): An echelon matrix that ...

echelon matrix (or **row echelon matrix**): A rectangular matrix that has three properties: (1) ... (2) ... (3) ....

elementary row operations: (1) ... (2) ... (3) ....

equivalent (linear) systems: Linear systems with ....

existence question: Asks, "Does ... exist?" or "Is ...?" Also, "Does ... exist for ...?"

free variable: Any variable in a linear system ....

Gaussian elimination: *See* row reduction algorithm.

general solution (of a linear system): A ... description of a solution set that expresses ....

inconsistent linear system: A linear system with ....

leading entry: The ... entry in a row of a matrix.

linear equation (in the variables $x_1, \ldots, x_n$): An equation that can be written in the form ....

linear system: A collection of one or more ... equations involving ....

matrix: ....

$m \times n$ matrix: A matrix with ....

overdetermined system: A system of equations with ....

pivot: A ... number that either is used ... or is ....

pivot column: A column that ....

pivot position: A position that ....

reduced echelon form (or **reduced row echelon form**, of a matrix): A ... matrix that is ....

reduced echelon matrix: A rectangular matrix in echelon form that has these additional properties: ....

row equivalent (matrices): Two matrices for which there exists ....

row reduced (matrix): A matrix that has been transformed ....

row reduction algorithm: A systematic method using ....

row replacement: An elementary row operation that ....

size (of a matrix): Two numbers ....

solution (of a linear system): ....

solution set: The set of ....

submatrix (of A): A matrix obtained by ....

system of linear equations (or a **linear system**): A collection of ....

underdetermined system: A system of equations with ....

uniqueness question: Asks, "If a solution of a system ...?"

2

Vector and Matrix Equations

2.1 VECTORS IN $\mathbb{R}^N$

As you work through this chapter and the next, your experience may resemble several walks through a village at different seasons of the year. The surroundings will be familiar, but the landscape will change. You will examine various mathematical concepts from several points of view, and a major problem will be to learn all the new terminology and the many connections between the concepts. In Chapter 5, you will see these ideas in a more abstract setting. Diligent work now will make the trip through Chapter 5 just another walk through the same village.

KEY IDEA

Do not be deceived by the rather simple beginning of Section 2.1. The important material on Span$\{\mathbf{v}_1, \ldots, \mathbf{v}_p\}$ will take time to digest. You cannot afford to postpone studying it. Exercises 15-26 are important. Each exercise involves an *existence* question about whether a certain vector equation has a solution. (You don't have to find the solution.) Note how the same basic question can be asked in several different ways.

STUDY NOTES

Develop the habit of reading the section carefully once or twice before looking at the Study Guide and before starting the exercises. (Don't just look at the pictures and examples! Important comments lurk in between.)

In nearly all of the text, a *scalar* is just a real number. By convention, scalars usually are written to the left of vectors, such as 5**v** or *c***v**, rather than **v**5 or **v***c*. To identify vectors in your lecture notes and homework, you can write underlined letters for vectors. (Some students write arrows above the letters, but that takes longer.)

Vectors must be the same size to be added or used in a linear combination. For instance, a vector in $\mathbb{R}^3$ cannot be added to a vector in $\mathbb{R}^2$.

SOLUTIONS TO EXERCISES

1. $\mathbf{u} = \begin{bmatrix} 3 \\ 2 \end{bmatrix}$, $\mathbf{v} = \begin{bmatrix} 2 \\ -1 \end{bmatrix}$. $\mathbf{u} + \mathbf{v} = \begin{bmatrix} 3 + 2 \\ 2 + (-1) \end{bmatrix} = \begin{bmatrix} 5 \\ 1 \end{bmatrix}$

$$\mathbf{u} - 2\mathbf{v} = \begin{bmatrix} 3 \\ 2 \end{bmatrix} - 2\begin{bmatrix} 2 \\ -1 \end{bmatrix} = \underbrace{\begin{bmatrix} 3 + (-2)(2) \\ 2 + (-2)(-1) \end{bmatrix}}_{\text{usually not written}} = \begin{bmatrix} 3 - 2(2) \\ 2 - 2(-1) \end{bmatrix} = \begin{bmatrix} -1 \\ 4 \end{bmatrix}$$

7. $\begin{bmatrix} s \\ 2 \end{bmatrix} = \begin{bmatrix} 5 \\ s - t \end{bmatrix}$. The equality of vectors yields: $\begin{array}{l} s = 5 \\ 2 = s - t \end{array}$

Substituting 5 for s in the second equation leads to $t = 5 - 2 = 3$.

13. The given system is already nicely aligned, with the coefficients of like variables arranged in columns.

$$\begin{aligned} 2x_1 - x_2 + 5x_3 &= 3 \\ x_1 - 8x_2 + 2x_3 &= 5 \\ 4x_2 - 4x_3 &= 5 \end{aligned} \qquad x_1\begin{bmatrix} 2 \\ 1 \\ 0 \end{bmatrix} + x_2\begin{bmatrix} -1 \\ -8 \\ 4 \end{bmatrix} + x_3\begin{bmatrix} 5 \\ 2 \\ -4 \end{bmatrix} = \begin{bmatrix} 3 \\ 5 \\ 5 \end{bmatrix}$$

19. Denote the columns of A by $\mathbf{a}_1$, $\mathbf{a}_2$, $\mathbf{a}_3$. To determine if $\mathbf{b}$ is a linear combination of these columns, use the boxed fact on page 44. Row reduce the augmented matrix:

$$[\mathbf{a}_1\ \ \mathbf{a}_2\ \ \mathbf{a}_3\ \ \mathbf{b}] = \begin{bmatrix} 1 & 0 & 2 & -5 \\ -2 & 1 & 0 & 11 \\ 2 & 1 & 8 & -7 \end{bmatrix} \sim \begin{bmatrix} 1 & 0 & 2 & -5 \\ 0 & 1 & 4 & 1 \\ 0 & 1 & 4 & 3 \end{bmatrix} \sim \begin{bmatrix} 1 & 0 & 2 & -5 \\ 0 & 1 & 4 & 1 \\ 0 & 0 & 0 & 2 \end{bmatrix}$$

The system for this augmented matrix is inconsistent, so $\mathbf{b}$ is *not* a linear combination of the columns of A.

25. a. There are only three vectors in the set $\{\mathbf{a}_1, \mathbf{a}_2, \mathbf{a}_3\}$, and $\mathbf{b}$ is not one of them.

b. There are infinitely many vectors in $W = \text{Span}\{\mathbf{a}_1, \mathbf{a}_2, \mathbf{a}_3\}$. To determine if $\mathbf{b}$ is in W, use the method of Exercise 19.

$$\begin{matrix} \begin{bmatrix} 1 & 0 & -4 & 4 \\ 0 & 1 & -2 & 1 \\ -2 & 3 & 3 & -4 \end{bmatrix} \\ \begin{matrix} \uparrow & \uparrow & \uparrow & \uparrow \\ \mathbf{a}_1 & \mathbf{a}_2 & \mathbf{a}_3 & \mathbf{b} \end{matrix} \end{matrix} \sim \begin{bmatrix} 1 & 0 & -4 & 4 \\ 0 & 1 & -2 & 1 \\ 0 & 3 & -5 & 4 \end{bmatrix} \sim \begin{bmatrix} 1 & 0 & -4 & 4 \\ 0 & 1 & -2 & 1 \\ 0 & 0 & 1 & 1 \end{bmatrix}$$

The system for this augmented matrix is consistent, so $\mathbf{b}$ is in W.

c. Are you reading the text? The answer is at the bottom of page 42.

MATLAB To access the data for Section 2.1, give the command **x2s1**. The data for Exercise 25, for example, consists of a matrix A, its columns **a1**, **a2**, **a3**, and the vector **b**. The command **M = [a1 a2 a3 b]** creates a matrix using the vectors as its columns. The same matrix is created by the command **M = [A b]**.

Each time you want data for a new exercise in Section 2.1, you need the command **x2s1**. After the first exercise, you can use the up-arrow (↑). This will make MATLAB scroll back through your old commands. You may be able to find "x2s1" faster than you can retype it. Press <Enter> to reuse the command.

2.2 THE EQUATION $A\mathbf{x} = \mathbf{b}$

You must read this section extremely carefully. The ideas, boxed statements, and theorems are absolutely fundamental for the rest of the text.

KEY IDEAS

The definition of $A\mathbf{x}$ as a linear combination of the columns of A will be used often. You should learn the definition in *words* as well as symbols. Note: $A\mathbf{x}$ is *not* written as $\mathbf{a}_1x_1 + \cdots + \mathbf{a}_nx_n$, because the scalar weights $x_1, \ldots, x_n$ usually appear to the *left* of the vectors.

You need to understand *why* Theorem 2 is true. That may take some time and effort. Example 2 should help, along with the proof. Theorem 2(c) can be restated as "The reduced echelon form of A has no row of zeros."

The phrase *logically equivalent* is explained in the statement of Theorem 2. This phrase is used with several statements in the same way that *if and only if*, or the symbol ⇔, is used between two statements. (See the Mathematical Note on page 1-9 in this Study Guide.) Saying that statements (a), (b), and (c) are logically equivalent means the same thing as saying that (a) ⇔ (b) and (a) ⇔ (c).

Key exercises are 17-28, 35, 36. Think about 35 and 36, even if they are not assigned.

Checkpoint: True or False? If an augmented matrix $[A \ \ \mathbf{b}]$ has a pivot position in every row, then the equation $A\mathbf{x} = \mathbf{b}$ is consistent.

Note: You should work a checkpoint problem when you see it, provided that you have already read the text at least once. Always *write* your answer before comparing it with the one I have written. The checkpoint answer will be at the end of the solutions to the exercises.

SOLUTIONS TO EXERCISES

1. The text has the solution. Exercises 1-6 are designed to help you learn the definition of $A\mathbf{x}$.

7. The solution is in the text. The goal of Exercises 7-16 is to help you learn Theorem 1. If a problem involves vectors, say $\mathbf{v}_1, \mathbf{v}_2, \mathbf{v}_3$, you can place the vectors into a matrix $[\mathbf{v}_1 \ \ \mathbf{v}_2 \ \ \mathbf{v}_3]$, if that is helpful. If a problem involves a matrix A, you can give names to the columns of A, say $\mathbf{a}_1, \mathbf{a}_2, \mathbf{a}_3$, and reformulate a matrix equation as a vector equation.

13. The given equation is in the form $A\mathbf{x} = \mathbf{b}$. The equivalent vector equation is in the text's answer. Note how the entries in the vector $\mathbf{x}$ are used as the weights in the linear combination of the columns of A.

17. To justify the answer, make an appropriate calculation, and then write a sentence that mentions an important fact described in this section.

19. The augmented matrix for $A\mathbf{x} = \mathbf{b}$ is $\begin{bmatrix} -3 & 1 & b_1 \\ 6 & -2 & b_2 \end{bmatrix}$. One row operation produces $\begin{bmatrix} -3 & 1 & b_1 \\ 0 & 0 & b_2 + 2b_1 \end{bmatrix}$, which shows that the equation $A\mathbf{x} = \mathbf{b}$ is not consistent unless $2b_1 + b_2 = 0$, that is, unless $b_2 = -2b_1$. The set of $\mathbf{b}$ for which the equation *is* consistent is a line through the origin—the set of all points (x_1, x_2) satisfying $x_2 = -2x_1$.

25. $A = \begin{bmatrix} 0 & 0 & 2 \\ 0 & -5 & 1 \\ 4 & 6 & -3 \end{bmatrix} \sim \begin{bmatrix} 4 & 6 & -3 \\ 0 & -5 & 1 \\ 0 & 0 & 2 \end{bmatrix}$, so the matrix has a pivot in each row. By Theorem 2, the columns of A span $\mathbb{R}^3$.

Study Tip: The answer shown here for Exercise 25 is the sort of answer expected for Exercises 23-28. A simple calculation is not enough. The phrase "*so the matrix has a pivot in each row*" is needed because it explains how Theorem 2 is used. On a test, you probably would not have to know the theorem number. It might be enough to say "*By a theorem*", instead of "*By Theorem 2*". (Check with your instructor.)

31. The text has the solution.

35. If A is $m \times n$ with more rows than columns, then A cannot possibly have a pivot in every row, so the three statements in Theorem 2 are all false. Thus, the equation $A\mathbf{x} = \mathbf{b}$ cannot be consistent for all $\mathbf{b}$ in $\mathbb{R}^m$.

Answer to Checkpoint: False. See page 51. If you missed this, you are not studying the text properly. You should read the text thoroughly *before* you look at the Study Guide and before you work on the exercises.

MATLAB To solve $A\mathbf{x} = \mathbf{b}$, row reduce the matrix **M = [A b]**. See the MATLAB notes in Section 1.1. The command **x = [5;3;-7]** creates a column vector **x** with entries 5,3,-7. Matrix-vector multiplication is **A*x**.

2.3 SOLUTION SETS OF LINEAR SYSTEMS

Many of the concepts and computations in linear algebra involve sets of vectors which are visualized geometrically as lines and planes. The most important examples of such sets are the solution sets of linear systems.

KEY IDEAS

Visualize the solution set of a homogeneous equation $A\mathbf{x} = \mathbf{0}$ as:

- the single point $\mathbf{0}$, when $A\mathbf{x} = \mathbf{0}$ has only the trivial solution,
- a line through $\mathbf{0}$, when $A\mathbf{x} = \mathbf{0}$ has one free variable,
- a plane through $\mathbf{0}$, when $A\mathbf{x} = \mathbf{0}$ has two free variables. (For more than two free variables, also use a plane through $\mathbf{0}$.)

For $\mathbf{b} \neq \mathbf{0}$, visualize the solution set of $A\mathbf{x} = \mathbf{b}$ as:

- empty, if $\mathbf{b}$ is not a linear combination of the columns of A,
- one nonzero point (vector), when $A\mathbf{x} = \mathbf{b}$ has a unique solution,
- a line not through $\mathbf{0}$, when $A\mathbf{x} = \mathbf{b}$ is consistent and has one free variable,
- a plane not through $\mathbf{0}$, when $A\mathbf{x} = \mathbf{b}$ is consistent and has two or more free variables.

The solution set of $A\mathbf{x} = \mathbf{b}$ is said to be described *implicitly*; the equation is a condition an $\mathbf{x}$ must satisfy in order to be in the set, yet the equation does not show how to find such an $\mathbf{x}$. When the solution set of $A\mathbf{x} = \mathbf{0}$ is written as Span$\{\mathbf{v}_1, \ldots, \mathbf{v}_p\}$, the set is said to be described *explicitly*; each element in the set is produced by forming a linear combination of $\mathbf{v}_1, \ldots, \mathbf{v}_p$.

A common explicit description of a set is an equation in *parametric vector form*. Examples are:

$\mathbf{x} = t\mathbf{v}$,	a line through $\mathbf{0}$ in the direction of $\mathbf{v}$,
$\mathbf{x} = \mathbf{p} + t\mathbf{v}$,	a line through $\mathbf{p}$ in the direction of $\mathbf{v}$,
$\mathbf{x} = x_2\mathbf{u} + x_3\mathbf{v}$,	a plane through $\mathbf{0}$, $\mathbf{u}$ and $\mathbf{v}$,
$\mathbf{x} = \mathbf{p} + x_2\mathbf{u} + x_3\mathbf{v}$,	a plane through $\mathbf{p}$ parallel to the plane whose equation is $\mathbf{x} = x_2\mathbf{u} + x_3\mathbf{v}$.

Such an equation describes a set explictly because the equation shows how to produce each **x** in the set.

Solving an equation $A\mathbf{x} = \mathbf{b}$ means to find an explicit description of the solution set. When this description is in parametric vector form, the parameters are the free variables from the system $A\mathbf{x} = \mathbf{b}$. Note that the number of free variables depends only on A, not on $\mathbf{b}$.

Theorem 4 and the paragraph following it are important. They describe how the solutions of $A\mathbf{x} = \mathbf{0}$ and $A\mathbf{x} = \mathbf{b}$ are related. See Figs. 6 and 7.

SOLUTIONS TO EXERCISES

1. The solution is in the text. The equations $\mathbf{x} = \mathbf{a} + t\mathbf{b}$ and $\mathbf{x} = \mathbf{b} + t\mathbf{a}$ are not the same. One line goes *through* **a** and the other is *parallel* to **a**. You can decide which is which by setting $t = 0$. The line passes through the point corresponding to $t = 0$.

7. This exercise illustrates a useful fact: Any homogeneous system of linear equations with more variables than there are equations must have at least one free variable and hence infinitely many nontrivial solutions.

Warning: The presence of free variables in a system of equations does not guarantee that there are infinitely many solutions. There may be *no* solution. (Of course, a homogeneous system always has a solution.)

13. Always use the *reduced* echelon form of an augmented matrix to find solutions of a system. See the text's discussion of back substitution on page 21.

$$\begin{bmatrix} 1 & -5 & 0 & 2 & 0 & -4 \\ 0 & 0 & 0 & 1 & 0 & -3 \\ 0 & 0 & 0 & 0 & 1 & 5 \\ 0 & 0 & 0 & 0 & 0 & 0 \end{bmatrix} \sim \begin{bmatrix} 1 & -5 & 0 & 0 & 0 & 2 \\ 0 & 0 & 0 & 1 & 0 & -3 \\ 0 & 0 & 0 & 0 & 1 & 5 \\ 0 & 0 & 0 & 0 & 0 & 0 \end{bmatrix}, \qquad \begin{aligned} x_1 - 5x_2 \qquad\qquad &= 2 \\ x_4 \quad &= -3 \\ x_5 &= 5 \end{aligned}$$

If you wrote something like this, then you made a common mistake. The matrix in the problem is a coefficient matrix, not an augmented matrix. You should row reduce $[A \;\; \mathbf{0}]$. The correct system of equations is

$$\begin{aligned} x_1 - 5x_2 \qquad\qquad + 2x_6 &= 0 \\ x_4 \qquad - 3x_6 &= 0 \\ x_5 + 5x_6 &= 0 \end{aligned}$$

Some students are not sure what to do with x_3. Some ignore it, others set it equal to zero. In fact, x_3 is free; there is no constraint on x_3 at all. So $x_1 = 5x_2 - 2x_6$, $x_4 = 3x_6$, $x_5 = -5x_6$, with x_2, x_3, x_6 free. The general solution is

$$\mathbf{x} = \begin{bmatrix} 5x_2 - 2x_6 \\ x_2 \\ x_3 \\ 3x_6 \\ -5x_6 \\ x_6 \end{bmatrix} = \begin{bmatrix} 5x_2 \\ x_2 \\ 0 \\ 0 \\ 0 \\ 0 \end{bmatrix} + \begin{bmatrix} 0 \\ 0 \\ x_3 \\ 0 \\ 0 \\ 0 \end{bmatrix} + \begin{bmatrix} -2x_6 \\ 0 \\ 0 \\ 3x_6 \\ -5x_6 \\ x_6 \end{bmatrix} = x_2 \underset{\uparrow \atop \mathbf{u}}{\begin{bmatrix} 5 \\ 1 \\ 0 \\ 0 \\ 0 \\ 0 \end{bmatrix}} + x_3 \underset{\uparrow \atop \mathbf{v}}{\begin{bmatrix} 0 \\ 0 \\ 1 \\ 0 \\ 0 \\ 0 \end{bmatrix}} + x_6 \underset{\uparrow \atop \mathbf{w}}{\begin{bmatrix} -2 \\ 0 \\ 0 \\ 3 \\ -5 \\ 1 \end{bmatrix}}$$

The solution set is the same as Span$\{\mathbf{u}, \mathbf{v}, \mathbf{w}\}$. Originally, the solution set was described implicitly, by a set of equations. Now the solution set is described explicitly, in parametric vector form.

Study Tip: When solving a system, identify (and perhaps circle) the basic variables. All other variables are free.

19. Row reduce the augmented matrix:

$$\begin{bmatrix} 1 & -3 & -2 & -5 \\ 0 & 1 & -1 & 4 \\ -2 & 3 & 7 & -2 \end{bmatrix} \sim \cdots \sim \begin{bmatrix} 1 & 0 & -5 & 7 \\ 0 & 1 & -1 & 4 \\ 0 & 0 & 0 & 0 \end{bmatrix}, \quad \begin{aligned} x_1 \quad - 5x_3 &= 7 \\ x_2 - x_3 &= 4 \\ 0 &= 0 \end{aligned}$$

So $x_1 = 7 + 5x_3$, $x_2 = 4 + x_3$, with x_3 free, and $\mathbf{x} = \begin{bmatrix} 7 \\ 4 \\ 0 \end{bmatrix} + x_3 \begin{bmatrix} 5 \\ 1 \\ 1 \end{bmatrix}$. The solution set is the line through $\begin{bmatrix} 7 \\ 4 \\ 0 \end{bmatrix}$, parallel to the solution set (a line) determined by the homogeneous system in Exercise 9.

Checkpoint: Let A be an $m \times n$ matrix. Answer True or False: If the solution set of $A\mathbf{x} = \mathbf{0}$ is a line through the origin in $\mathbb{R}^3$ and if $\mathbf{b} \neq \mathbf{0}$, then the solution set of $A\mathbf{x} = \mathbf{b}$ is a line not through the origin.

25. Suppose $\mathbf{p}$ satisfies $A\mathbf{x} = \mathbf{b}$. Then $A\mathbf{p} = \mathbf{b}$. Theorem 4 says that the solution set of $A\mathbf{x} = \mathbf{b}$ equals the set $S = \{\mathbf{w} : \mathbf{w} = \mathbf{p} + \mathbf{v}_h$ for some $\mathbf{v}_h$ such that $A\mathbf{v}_h = \mathbf{0}\}$. There are two things to prove: (a) every vector in S satisfies $A\mathbf{x} = \mathbf{b}$, (b) every vector that satisfies $A\mathbf{x} = \mathbf{b}$ is in S.

a. Let $\mathbf{w}$ have the form $\mathbf{w} = \mathbf{p} + \mathbf{v}_h$, where $A\mathbf{v}_h = \mathbf{0}$. Then

$$A\mathbf{w} = A(\mathbf{p} + \mathbf{v}_h) = A\mathbf{p} + A\mathbf{v}_h \qquad \text{By Theorem 3(a) in Section 2.2}$$
$$= \mathbf{b} + \mathbf{0} = \mathbf{b}$$

So every vector of the form $\mathbf{p} + \mathbf{v}_h$ satisfies $A\mathbf{x} = \mathbf{b}$.

b. Now let $\mathbf{w}$ be any solution of $A\mathbf{x} = \mathbf{b}$, and set $\mathbf{v}_h = \mathbf{w} - \mathbf{p}$. Then

$$A\mathbf{v}_h = A(\mathbf{w} - \mathbf{p}) = A\mathbf{w} - A\mathbf{p} = \mathbf{b} - \mathbf{b} = \mathbf{0}$$

So $\mathbf{v}_h$ satisfies $A\mathbf{x} = \mathbf{0}$. Thus every solution of $A\mathbf{x} = \mathbf{b}$ has the form $\mathbf{w} = \mathbf{p} + \mathbf{v}_h$.

31. A is a 3×2 matrix with two pivot positions. (a) Since A has a pivot position in each column, each variable in $A\mathbf{x} = \mathbf{0}$ is a basic variable. So there are no free variables, and $A\mathbf{x} = \mathbf{0}$ has no nontrivial solution. (b) A cannot have a pivot position in each of its three rows, so the equation $A\mathbf{x} = \mathbf{b}$ does not have a solution for *every possible* $\mathbf{b}$, by Theorem 2 in Section 2.2.

Note: The term *possible* here means that of course we only consider $\mathbf{b}$ in $\mathbb{R}^3$, because A has 3 rows. The context of the problem determines what vectors are possible candidates for $\mathbf{b}$.

Answer to Checkpoint: False. The solution set could be empty. (See the paragraph in the text preceding Theorem 4.) Suppose $A = \begin{bmatrix} 1 & 2 \\ 1 & 2 \end{bmatrix}$, $\mathbf{v} = \begin{bmatrix} -2 \\ 1 \end{bmatrix}$, and $\mathbf{b} = \begin{bmatrix} 5 \\ h \end{bmatrix}$. Then the general solution of $A\mathbf{x} = \mathbf{0}$ is the line $\mathbf{x} = t\mathbf{v}$, but if $h \neq 5$, the solution set of $A\mathbf{x} = \mathbf{b}$ is empty. (You are not expected to furnish such example at this point in the course.)

MATLAB The command **zeros(m,n)** creates an $m \times n$ matrix of zeros. When solving an equation $A\mathbf{x} = \mathbf{0}$, create an augmented matrix:

```
M = [A  zeros(m,1)]        m is the number of rows in A.
```

Then use **replace**, **swap**, and **scale** to completely row reduce M.

The entry in row r and column c of a matrix M is denoted by $M(r,c)$. If the number stored in $M(r,c)$ is displayed with a decimal point, then the displayed value may be accurate to only about five digits. In this case, use the *symbol* $M(r,c)$ instead of the displayed value in calculations.

For instance, if you want to scale row r of M to change the value of $M(r,c)$ to 1, and if $M(r,c)$ is displayed as a decimal, use the command

```
M = scale(A,r,1/M(r,c))
```

If you want to use a pivot entry $M(s,c)$ to change $M(r,c)$ to 0, enter the commands

```
m = -M(r,c)/M(s,c)       The multiple of row s to be added to row r
M = replace(M,r,m,s)     Adds m times row s to row r
```

Or, you can use just one command: **M = replace(M,r,-M(r,c)/M(s,c),s)**.

Once you have entered a command, you can recall it by pressing the up-arrow (↑) key one or more times. If needed, edit the command line by typing in changes. Press <Return> to reuse the (edited) command.

2.4 LINEAR INDEPENDENCE

This section is as important as Section 2.2 and must be studied just as carefully. Full understanding of the concepts will take time, so get started on the section now.

KEY IDEAS

Figures 1 and 2, along with Theorem 5, will help you understand the nature of a linearly dependent set. (Fig. 2 applies only when **u** and **v** are independent.) But you must also learn the *definitions* of linear dependence and linear independence, word for word! Many theoretical problems involving a linearly dependent set are treated by the definition, because it provides an equation (the dependence equation) with which to work. (See the proof of Theorem 5.)

The box before Example 2 contains a very useful fact. Any time you need to study the linear independence of a set of p vectors in $\mathbb{R}^n$, you can always form an $n \times p$ matrix A with those vectors as columns and then study the matrix equation $A\mathbf{x} = \mathbf{0}$. This is not the only method, however. Stay alert for three special situations:

- A set of two vectors. Always check this by inspection; don't waste time on row reduction of $[A \quad \mathbf{0}]$. The set is linearly independent if neither of the vectors is a multiple of the other. (For brevity, I often say that "the vectors are not multiples.") See Example 3.
- A set that contains too many vectors, that is, more vectors than entries in the vectors; the columns of a short, fat matrix. Theorem 6.
- A set that contains the zero vector. Theorem 7.

The most common mistake students make when checking a set of three or more vectors for independence is to think they only have to verify that no vector is a multiple of one of the other vectors. That's wrong! Study Example 5 and Figure 4.

SOLUTIONS TO EXERCISES

1. Let $\mathbf{u} = \begin{bmatrix} 3 \\ 0 \\ 0 \end{bmatrix}$, $\mathbf{v} = \begin{bmatrix} -3 \\ 2 \\ 3 \end{bmatrix}$, $\mathbf{w} = \begin{bmatrix} 6 \\ 4 \\ 0 \end{bmatrix}$. To test the linear independence of $\{\mathbf{u}, \mathbf{v}, \mathbf{w}\}$, use an augmented matrix to study the solution set of

$$x_1\mathbf{u} + x_2\mathbf{v} + x_3\mathbf{w} = \mathbf{0} \qquad (*)$$

Since $\begin{bmatrix} 3 & -3 & 6 & 0 \\ 0 & 2 & 4 & 0 \\ 0 & 3 & 0 & 0 \end{bmatrix} \sim \begin{bmatrix} 3 & -3 & 6 & 0 \\ 0 & 2 & 4 & 0 \\ 0 & 0 & -6 & 0 \end{bmatrix}$, there are three basic variables, no free variables, and so (*) has *only* the trivial solution. The vectors are linearly independent.

Warning: Whenever you study a homogeneous equation, you may be tempted to omit the augmented column of zeros because it never changes under row operations. I urge you to keep the zeros, to avoid possibly misinterpreting your own calculations. In Exercise 1, if you wrote

$$\begin{bmatrix} 3 & -3 & 6 \\ 0 & 2 & 4 \\ 0 & 3 & 0 \end{bmatrix} \sim \begin{bmatrix} 3 & -3 & 6 \\ 0 & 2 & 4 \\ 0 & 0 & -6 \end{bmatrix}$$

you might conclude that "the system is inconsistent" and then go on to make some crazy statement about linear dependence or independence. Don't laugh. I have seen this happen on exams. A more common error occurs in a problem like Exercise 11. In that exercise, if you write

$$\begin{bmatrix} 1 & 1 & 0 & 4 \\ -1 & 0 & 3 & -1 \\ 0 & -2 & 1 & 1 \\ 1 & 0 & -1 & 3 \end{bmatrix} \sim \cdots \sim \begin{bmatrix} 1 & 1 & 0 & 4 \\ 0 & 1 & 3 & 3 \\ 0 & 0 & 7 & 7 \\ 0 & 0 & 0 & 0 \end{bmatrix}$$

you might conclude that "the system has a unique solution", and so the vectors are linearly independent. (Actually, the four vectors are linearly dependent.) In both cases, the error is to misinterpret your matrix as an augmented matrix.

7. Study the equation $A\mathbf{x} = \mathbf{0}$. You could row reduce $\begin{bmatrix} 1 & 3 & -2 & 0 & 0 \\ 3 & 10 & -7 & 1 & 0 \\ -5 & -5 & 3 & 7 & 0 \end{bmatrix}$,

but that would be a waste of time. There are only three rows, so there are at most three pivot positions. At least one of the four variables must be free. So the equation $A\mathbf{x} = \mathbf{0}$ has a nontrivial solution and the columns of A are linearly dependent. (If you know Theorem 6, you can even omit most of this discussion.)

Checkpoint: What is wrong with the following statement (which is similar to statements I frequently see in student work)?

The vectors $\begin{bmatrix} 3 \\ -1 \end{bmatrix}$, $\begin{bmatrix} 2 \\ 8 \end{bmatrix}$, $\begin{bmatrix} -5 \\ 3 \end{bmatrix}$, $\begin{bmatrix} 7 \\ -4 \end{bmatrix}$ are linearly dependent "because there is a free variable", or "because there are more variables than equations".

13. $\mathbf{v}_1 = \begin{bmatrix} 1 \\ 3 \\ -2 \end{bmatrix}$, $\mathbf{v}_2 = \begin{bmatrix} -2 \\ -6 \\ 4 \end{bmatrix}$, $\mathbf{v}_3 = \begin{bmatrix} 1 \\ 2 \\ h \end{bmatrix}$

a. For $\mathbf{v}_3$ to be in Span$\{\mathbf{v}_1, \mathbf{v}_2\}$, the system $x_1\mathbf{v}_1 + x_2\mathbf{v}_2 = \mathbf{v}_3$ must be consistent. To find out if this is true, row reduce $[\mathbf{v}_1 \;\; \mathbf{v}_2 \;\; \mathbf{v}_3]$, considered as an augmented matrix.

$$\begin{bmatrix} 1 & -2 & 1 \\ 3 & -6 & 2 \\ -2 & 4 & h \end{bmatrix} \sim \begin{bmatrix} 1 & -2 & 1 \\ 0 & 0 & -1 \\ -2 & 4 & h \end{bmatrix}$$

No further work is needed. The second equation, $0x_1 + 0x_2 = -1$, is impossible, so the original system is inconsistent. Thus $\mathbf{v}_3$ is not in Span$\{\mathbf{v}_1, \mathbf{v}_2\}$ for any value of h.

b. For $\{\mathbf{v}_1, \mathbf{v}_2, \mathbf{v}_3\}$ to be linearly independent, the system $x_1\mathbf{v}_1 + x_2\mathbf{v}_2 + x_3\mathbf{v}_3 = \mathbf{0}$ should have only the trivial solution. Row reduce the associated augmented matrix $[\mathbf{v}_1 \;\; \mathbf{v}_2 \;\; \mathbf{v}_3 \;\; \mathbf{0}]$:

$$\begin{bmatrix} 1 & -2 & 1 & 0 \\ 3 & -6 & 2 & 0 \\ -2 & 4 & h & 0 \end{bmatrix} \sim \begin{bmatrix} 1 & -2 & 1 & 0 \\ 0 & 0 & -1 & 0 \\ 0 & 0 & h+2 & 0 \end{bmatrix} \sim \begin{bmatrix} 1 & -2 & 1 & 0 \\ 0 & 0 & -1 & 0 \\ 0 & 0 & 0 & 0 \end{bmatrix}$$

For every value of h, x_2 is a free variable, and so the homogeneous system has a nontrivial solution. Thus $\{\mathbf{v}_1, \mathbf{v}_2, \mathbf{v}_3\}$ is linearly dependent for all h. (Note: You can avoid all calculation if you happen to notice that $\mathbf{v}_2 = -2\mathbf{v}_1$, and then mention Theorem 5. In general, however, you should not spend much time looking for special cases such as this.)

Warning: Exercise 13 and Practice Problem 3 emphasize that to check whether a set such as $\{\mathbf{v}_1, \mathbf{v}_2, \mathbf{v}_3\}$ is linearly independent, it is *not* wise to check instead whether $\mathbf{v}_3$ is a linear combination of $\mathbf{v}_1$ and $\mathbf{v}_2$.

19. The set $\begin{bmatrix} 5 \\ 5 \end{bmatrix}, \begin{bmatrix} 6 \\ 1 \end{bmatrix}, \begin{bmatrix} 2 \\ 4 \end{bmatrix}, \begin{bmatrix} 3 \\ 6 \end{bmatrix}$ is obviously linearly dependent, by Theorem 6, because there are more vectors (4) than entries in the vectors. (On a test, you would probably not have to know the theorem number.)

25. The answer in the text is correct because $A\mathbf{x} = \mathbf{0}$ has only the trivial solution if and only if the columns of A are linearly independent. If A has only two nonzero columns, they are linearly independent if and only if they are *not* multiples.

Examples: $A = \begin{bmatrix} 1 & 0 \\ 0 & 1 \\ 0 & 1 \end{bmatrix}$, $B = \begin{bmatrix} 2 & 1 \\ 2 & \\ 2 & \end{bmatrix}$

31. The text gives the answer. Note that if two of the vectors, say $\mathbf{v}_1$ and $\mathbf{v}_2$, are equal, then in a technical sense, the set $\{\mathbf{v}_1, \mathbf{v}_2, \mathbf{v}_3\}$ contains only two distinct vectors. Of course, a set of only two vectors in $\mathbb{R}^2$ could be linearly independent. However, we regard the notation "$\mathbf{v}_1, \mathbf{v}_2, \mathbf{v}_3$" as a list of three separate (but not necessarily distinct) vectors and we consider the equation $x_1\mathbf{v}_1 + x_2\mathbf{v}_2 + x_3\mathbf{v}_3 = \mathbf{0}$. Such an equation, with vectors in $\mathbb{R}^2$, obviously has a nontrivial solution. Similarly, if we write $A = [\mathbf{v}_1 \ \ \mathbf{v}_2 \ \ \mathbf{v}_3]$ (with $\mathbf{v}_1, \mathbf{v}_2, \mathbf{v}_3$ in $\mathbb{R}^2$), then the columns of A are linearly dependent, even though two of the columns might be equal.

37. If for all $\mathbf{b}$, the equation $A\mathbf{x} = \mathbf{b}$ has at most one solution, then take $\mathbf{b} = \mathbf{0}$, and conclude that the equation $A\mathbf{x} = \mathbf{0}$ has at most one solution. Of course the trivial solution is a solution, so it is the only solution. Thus the columns of A are linearly independent.

Answer to Checkpoint: The set of four vectors contains only vectors, no variables of any kind, and no equations. It makes no sense to talk about the variables in a set of vectors. Variables appear in an equation. One cannot assume that the writer of the statement has any idea of the appropriate equation. If you want to give an explanation involving variables, then you must specify the equation. One correct answer is: the vectors are linearly dependent because the equation $x_1\begin{bmatrix}3\\-1\end{bmatrix} + x_2\begin{bmatrix}2\\8\end{bmatrix} + x_3\begin{bmatrix}-5\\3\end{bmatrix} + x_4\begin{bmatrix}7\\-4\end{bmatrix} = \begin{bmatrix}0\\0\end{bmatrix}$ necessarily has a free variable.

Appendix: Mastering Linear Algebra Concepts

Begin by reviewing the "Steps to Success in Linear Algebra" on pages v-vi. Your mastery of linear algebra will be influenced greatly by how well you follow those steps.

To really understand a key concept, you need to form an image in your mind that consists of the basic definition together with many related ideas. The list below uses **linear dependence** as an example, but your goal for each concept in the course should be to mold many ideas into a single mental image, with each part immediately available for use as needed:

- geometric interpretations, — **Figures 1,2,4**
- equivalent definition(s), — **Theorem 5**
- special cases, — **Theorems 6,7, box on p. 66, Examples 3,5,6**
- examples and "counterexamples", — **Figures 1,2,3,4, Exercises 13-18, 29-34**
- algorithms or typical computations, — **Examples 1,2, Exercises 1-12**
- connections with other concepts. — **Box on p. 65, Examples 2,4, Exercises 35-37**

2.5 INTRODUCTION TO LINEAR TRANSFORMATIONS

Linear transformations are important both for the theory and the applications of linear algebra. You will see both uses in a variety of settings throughout the text. The graphical descriptions in this section will be augmented in a later section on computer graphics.

STUDY NOTES

Viewing the correspondence from a vector **x** to a vector $A\mathbf{x}$ as a mapping provides a dynamic interpretation of matrix-vector multiplication and a new way to understand the equation $A\mathbf{x} = \mathbf{b}$. Using the language of computer science, we can describe a matrix in two ways—as a data structure (a rectangular array of numbers) and as a program (a prescription for transforming vectors). Strictly speaking, however, the actual linear transformation is the function or mapping $\mathbf{x} \mapsto A\mathbf{x}$ rather than just A itself.

Here is a way to visualize a matrix acting as a linear transformation. The entries in the input vector **x** are assigned as weights that multiply the corresponding columns of A, then the resulting weighted columns are added together to produce the output vector **b**.

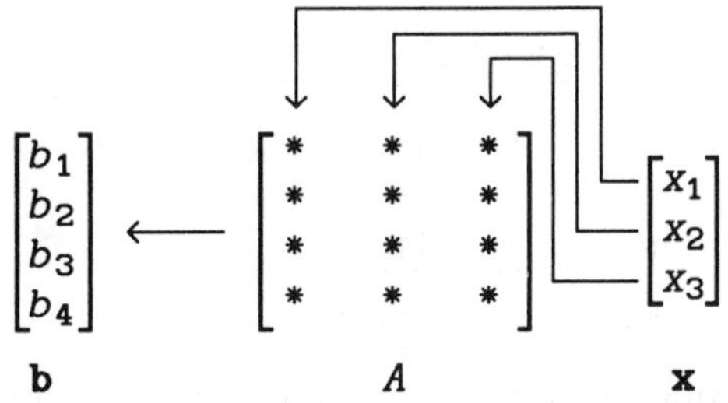

As you learn the definition of a linear transformation T, don't forget the crucial phrases "for all **u** and **v** in the domain of T" and "for all **u** and all scalars c." The mapping T defined by $T(x_1, x_2) = (|x_2|, |x_1|)$ is *not* a linear mapping, and yet T satisfies the linearity properties for *some* vectors in its domain and *some* scalars.

SOLUTIONS TO EXERCISES

1. $A = \begin{bmatrix} 3 & 0 \\ 0 & 3 \end{bmatrix}$. $T\mathbf{u} = \begin{bmatrix} 3 & 0 \\ 0 & 3 \end{bmatrix}\begin{bmatrix} 1 \\ 5 \end{bmatrix} = \begin{bmatrix} 3 \\ 15 \end{bmatrix}$, $T\mathbf{v} = \begin{bmatrix} 3 & 0 \\ 0 & 3 \end{bmatrix}\begin{bmatrix} -4 \\ -1 \end{bmatrix} = \begin{bmatrix} -12 \\ -3 \end{bmatrix}$.
Can you show that T is a dilation transformation?

7. If A is 7×5, then **x** needs to be in $\mathbb{R}^5$ for $A\mathbf{x}$ to be defined. Since $A\mathbf{x}$ is a linear combination of the columns of A and each column has 7 entries, $A\mathbf{x}$ is in $\mathbb{R}^7$. If $T(\mathbf{x}) = A\mathbf{x}$, then T maps $\mathbb{R}^a$ into $\mathbb{R}^b$, where $a = 5$ and $b = 7$.

13. a.

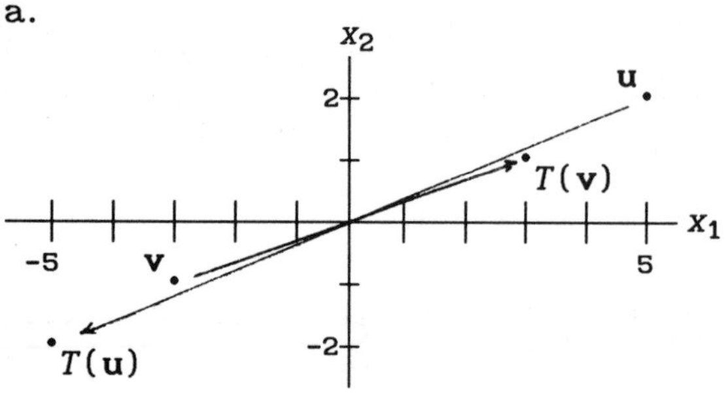

b. A reflection through the origin; two other good answers are: a a rotation of $\pm\pi$ radians about the origin.

19. By linearity, since $\mathbf{u}$ maps onto $\begin{bmatrix}2\\0\end{bmatrix}$, $2\mathbf{u}$ maps onto $2\begin{bmatrix}2\\0\end{bmatrix} = \begin{bmatrix}4\\0\end{bmatrix}$. Likewise, since $\mathbf{v}$ maps onto $\begin{bmatrix}1\\-4\end{bmatrix}$, $3\mathbf{v}$ maps onto $3\begin{bmatrix}1\\-4\end{bmatrix} = \begin{bmatrix}3\\-12\end{bmatrix}$. Finally, $T(2\mathbf{u} + 3\mathbf{v})$ is the sum of the images of $2\mathbf{u}$ and $3\mathbf{v}$, namely, $\begin{bmatrix}7\\-12\end{bmatrix}$.

23. A point $\mathbf{x}$ on the line through $\mathbf{p}$ in the direction of $\mathbf{v}$ satisfies the parametric equation $\mathbf{x} = \mathbf{p} + t\mathbf{v}$. By linearity, the image $T(\mathbf{x})$ satisfies the parametric equation

$$T(\mathbf{x}) = T(\mathbf{p} + t\mathbf{v}) = T(\mathbf{p}) + tT(\mathbf{v}) \qquad (*)$$

If $T(\mathbf{v}) = \mathbf{0}$, then $T(\mathbf{x}) = T(\mathbf{p})$ for all values of t, and the image of the original line is just a single point. Otherwise, (*) is the parametric equation of a line through $T(\mathbf{p})$ in the direction of $T(\mathbf{v})$.

Study Tip: Exercise 25 is important, because it will help you to connect the concepts of linear dependence and linear transformation. Be sure to *try* the exercise first, before looking in the answer section of the text. Don't feel bad if you need to peek at the hint there. Only my best students can do this problem unaided. Once you have seen the hint, try hard to construct the desired explanation without consulting the solution I have written below. Don't give up for at least ten minutes. Reread the definitions of linear dependence and linear transformation, if necessary.

After you have written your best attempt at an explanation, check it against the Study Guide solution. Also, study the strategy there of how I found the solution. Even if your attempt is quite unsatisfactory, the time spent on this problem is worthwhile, because you will learn more from the solution here.

25. To help you use this Study Guide properly, I have hidden the solution at the end of the solutions for Section 2.6. *Do not look there until you have followed the instructions above.*

2.6 THE MATRIX OF A LINEAR TRANSFORMATION

This section justifies the assertion in Section 2.5 that every linear transformation from $\mathbb{R}^n$ to $\mathbb{R}^m$ is a matrix transformation. Linear transformations on general vector spaces will be discussed in Chapters 5 and 6.

KEY IDEAS

A linear transformation $T:\mathbb{R}^n \to \mathbb{R}^m$ is completely determined by what it does to the columns of I_n. The jth column of the standard matrix for T is $T(\mathbf{e}_j)$, where $\mathbf{e}_j$ is the jth column of I_n.

There are two ways to compute the standard matrix A. Either compute $T(\mathbf{e}_1), \ldots, T(\mathbf{e}_n)$, which is easy to do when T is described geometrically, as in Exercises 1-14, or fill in the entries of A by inspection, which is easy to do when T is described by a formula, as in Exercises 15-24.

Existence and uniqueness questions about the mapping $\mathbf{x} \mapsto A\mathbf{x}$ are determined by properties of A. You should know how this works. The proof of Theorem 9 also applies to linear transformations on the general vector spaces in Chapter 5. Here is a shorter proof that applies only to matrix transformations.

> Let A be the standard matrix of T. Then T is one-to-one if and only if the equation $A\mathbf{x} = \mathbf{b}$ has at most one solution for each $\mathbf{b}$. This happens if and only if every column of A is a pivot column which, in turn, happens if and only if $A\mathbf{x} = \mathbf{0}$ has only the trivial solution.

SOLUTIONS TO EXERCISES

1. The columns of the standard matrix A of T are the images of $\mathbf{e}_1$ and $\mathbf{e}_2$. Since $T(\mathbf{e}_1) = \begin{bmatrix} 4 \\ -1 \\ 2 \end{bmatrix}$ and $T(\mathbf{e}_2) = \begin{bmatrix} -5 \\ 3 \\ -6 \end{bmatrix}$, we have $A = \begin{bmatrix} 4 & -5 \\ -1 & 3 \\ 2 & -6 \end{bmatrix}$.

7. $\mathbf{e}_1$ "reflects" onto itself, while $\mathbf{e}_2$ reflects onto $-\mathbf{e}_2$. So

$$A = [T(\mathbf{e}_1) \quad T(\mathbf{e}_2)] = [\mathbf{e}_1 \quad -\mathbf{e}_2] = \begin{bmatrix} 1 & 0 \\ 0 & -1 \end{bmatrix}$$

13. $T(\mathbf{e}_1) = \mathbf{e}_1 + 2\mathbf{e}_2 = \begin{bmatrix} 1 \\ 2 \end{bmatrix}$, $T(\mathbf{e}_2) = \mathbf{e}_2 = \begin{bmatrix} 0 \\ 1 \end{bmatrix}$; so $A = \begin{bmatrix} 1 & 0 \\ 2 & 1 \end{bmatrix}$.

Checkpoint 1. I recently put the following problem on an exam. There are no exercises like this in the text, but if you understand Theorem 8, you should be able to do it. Try it. The answer is at the end of the section.

Let $T:\mathbb{R}^2 \to \mathbb{R}^2$ be the mapping that first rotates points clockwise through $\pi/2$ radians about the origin, and then reflects the resulting points through the vertical x_2-axis. Assuming that T is a linear transformation, find its standard matrix A.

Checkpoint 2: Use an idea of this section to explain why the linear transformation T that reflects points in $\mathbb{R}^2$ though the origin, i.e., $(x_1, x_2) \mapsto (-x_1, -x_2)$ is the same as the linear transformation R that rotates points about the origin in $\mathbb{R}^2$ through π radians.

19. Use the method of the practice problem:

$$\begin{bmatrix} ? & ? & ? \\ ? & ? & ? \end{bmatrix}\begin{bmatrix} x_1 \\ x_2 \\ x_3 \end{bmatrix} = \begin{bmatrix} 3x_2 - x_3 \\ x_1 + 4x_2 + x_3 \end{bmatrix}, \quad \text{so} \quad A = \begin{bmatrix} 0 & 3 & -1 \\ 1 & 4 & 1 \end{bmatrix}$$

Study Tip: When T is described by a formula as in Exercises 15-24, you can use the method of Exercise 19 to find an A such that $T(\mathbf{x}) = A\mathbf{x}$, *provided* that T is a linear transformation. (Finding A *proves* that T is linear.) If you can't find the matrix, T is probably *not* a linear transformation. To show that such a T is not linear, you have to find either two vectors $\mathbf{u}$ and $\mathbf{v}$ such that $T(\mathbf{u} + \mathbf{v}) \neq T(\mathbf{u}) + T(\mathbf{v})$ or a vector $\mathbf{u}$ and a scalar c such that $T(c\mathbf{u}) \neq cT(\mathbf{u})$.

Note: The text does not give you practice determining whether a transformation is linear because the time needed to develop this skill would have to be taken away from some other topic. If you are expected to have this skill, you will need some exercises. Check with your instructor.

25. Row reduce the standard matrix A of the transformation T in Exercise 3:

$$A = \begin{bmatrix} 1 & -2 & 3 \\ 4 & 9 & -8 \end{bmatrix} \sim \begin{bmatrix} 1 & -2 & 3 \\ 0 & 17 & -20 \end{bmatrix}$$

Since the third column of A is not a pivot column, any equation of the form $A\mathbf{x} = \mathbf{b}$ will have x_3 as a free variable. So T is not one-to-one.

Another way to show that T is not one-to-one is to look at A and see that its columns are obviously linearly dependent (why?), and then use Theorem 10(b).

31. *T is one-to-one if and only if A has n pivot columns.* This statement follows by combining Theorem 10(b) with the statement in Exercise 36 of Section 2.4.

33. Define $T:\mathbb{R}^n \to \mathbb{R}^m$ by $T(\mathbf{x}) = B\mathbf{x}$ for some $m \times n$ matrix B, and let A be the standard matrix for T. By definition, $A = [T(\mathbf{e}_1) \ \cdots \ T(\mathbf{e}_n)]$, where $\mathbf{e}_j$ is the jth column of I_n. However, by matrix-vector mulitplication, $T(\mathbf{e}_j) = B\mathbf{e}_j = \mathbf{b}_j$, the jth column of B. So $A = [\mathbf{b}_1 \ \cdots \ \mathbf{b}_n] = B$.

25. *(This solution is for Section 2.5.) To construct the proof, first write in mathematical terms what is given.*

Since $\{\mathbf{v}_1, \mathbf{v}_2, \mathbf{v}_3\}$ is linearly dependent, there exist scalars c_1, c_2, c_3, not all zero, such that

$$c_1\mathbf{v}_1 + c_2\mathbf{v}_2 + c_3\mathbf{v}_3 = \mathbf{0} \qquad (*)$$

Next, think about what you must prove. In this problem, to prove that the image points are linearly dependent, you need a dependence relation among $T(\mathbf{v}_1)$, $T(\mathbf{v}_2)$, and $T(\mathbf{v}_3)$. That fact suggests the next step.

Apply T to both sides of (*) and use linearity of T, obtaining

$$T(c_1\mathbf{v}_1 + c_2\mathbf{v}_2 + c_3\mathbf{v}_3) = T(\mathbf{0})$$

and

$$c_1T(\mathbf{v}_1) + c_2T(\mathbf{v}_2) + c_3T(\mathbf{v}_3) = \mathbf{0}$$

Since not all the weights are zero, $\{T(\mathbf{v}_1), T(\mathbf{v}_2), T(\mathbf{v}_3)\}$ is a linearly dependent set. This completes the proof.

Study Tip: Analyze the strategy above for solving Exercise 25 (in Section 2.5). This approach will work later in a variety of situations.

Answers to Checkpoints:

1. $\mathbf{e}_1$ rotates to $\mathbf{e}_2$ and is then unchanged by the reflection; $\mathbf{e}_2$ rotates to $-\mathbf{e}_1$ and is then reflected to $+\mathbf{e}_1$. So $T(\mathbf{e}_1) = \mathbf{e}_2$, $T(\mathbf{e}_2) = \mathbf{e}_1$, and $A = [\mathbf{e}_2 \ \ \mathbf{e}_1] = \begin{bmatrix} 0 & 1 \\ 1 & 0 \end{bmatrix}$.

2. The reflection T has the property that $T(\mathbf{e}_1) = -\mathbf{e}_1$ and $T(\mathbf{e}_2) = -\mathbf{e}_2$, while the rotation R has the property $R(\mathbf{e}_1) = -\mathbf{e}_1$ and $R(\mathbf{e}_2) = -\mathbf{e}_2$. Since a linear transformation is completely determined by what it does to the columns $\mathbf{e}_1$ and $\mathbf{e}_2$ of the identity matrix, T and R must be the same transformation. (You could also explain this by observing that T and R have the same standard matrix, namely, $[-\mathbf{e}_1 \ \ -\mathbf{e}_2]$.)

2.7 APPLICATIONS TO NUTRITION AND POPULATION MOVEMENT

The application to nutrition relates to the problem described at the beginning of the chapter. The discussion of population movement introduces the concept of a difference equation, which will be studied in greater detail in Sections 5.8, 5.9, and 6.6.

KEY IDEAS

Nutrition Problem: In some applied problems such as the nutrition problem considered here, the data are already organized naturally in a manner that leads to a vector equation of the type we have discussed. The steps to the solution in this case may be diagrammed as follows:

Applied Problem → Vector Equation → Augmented Matrix → Solution Set → Answer to Problem

The discussion in Section 1.3 about an economic model could have used this approach if vector notation had been available. Another example is in Section 3.7.

The nutrition problem leads naturally into linear programming, a subject that uses linear algebra and has applications in agriculture, business, engineering, and other areas. In the 1950's and 1960's, one of the most common applications of linear algebra (measured in millions of dollars per year for computer time) was to linear programming problems. Such problems are still of great importance in operations research and management science. The following reference gives an entertaining introduction to linear programming. Matrix notation is used in the appendix (pp. 127-152).

> Saul I. Gass, *An Illustrated Guide to Linear Programming*, New York: McGraw-Hill, 1970. Republished by Dover Publications, 1990.

Population Movement: The entries in each *column* of the migration matrix must sum to one because the (decimal) fractions in a column account for the entire population in one region. A certain fraction of the population in a region remains in (or moves within) that region, and other fractions move elsewhere. (In Example 2 there are only two regions, but in Exercise 8 there are three regions.)

Warning: The order of the entries in a column of a migration matrix must match the order of the columns. For instance, if the first column concerns the population in the city, then the first entry in *each* column must be the fraction of a population that moves to (or remains in) the city.

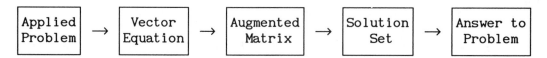

SOLUTIONS TO EXERCISES

1. a. If x_1 is the number of servings of Cheerios and x_2 is the number of servings of 100% Natural Cereal, then x_1 and x_2 should satisfy

$$x_1\begin{bmatrix}\text{nutrients}\\ \text{per serving}\\ \text{of Cheerios}\end{bmatrix} + x_2\begin{bmatrix}\text{nutrients}\\ \text{per serving of}\\ \text{100\% Natural}\end{bmatrix} = \begin{bmatrix}\text{list of}\\ \text{nutrients}\\ \text{desired}\end{bmatrix}$$

That is,

$$x_1\begin{bmatrix}110\\4\\20\\2\end{bmatrix} + x_2\begin{bmatrix}130\\3\\18\\5\end{bmatrix} = \begin{bmatrix}295\\9\\48\\8\end{bmatrix}$$

b. The equivalent matrix equation is $\begin{bmatrix}110 & 130\\4 & 3\\20 & 18\\2 & 5\end{bmatrix}\begin{bmatrix}x_1\\x_2\end{bmatrix} = \begin{bmatrix}295\\9\\48\\8\end{bmatrix}$. To solve this, row reduce the augmented matrix for this equation.

$$\begin{bmatrix}110 & 130 & 295\\4 & 3 & 9\\20 & 18 & 48\\2 & 5 & 8\end{bmatrix} \sim \begin{bmatrix}2 & 5 & 8\\4 & 3 & 9\\20 & 18 & 48\\110 & 130 & 295\end{bmatrix} \sim \begin{bmatrix}1 & 2.5 & 4\\4 & 3 & 9\\10 & 9 & 24\\110 & 130 & 295\end{bmatrix}$$

$$\sim \begin{bmatrix}1 & 2.5 & 4\\0 & -7 & -7\\0 & -16 & -16\\0 & -145 & -145\end{bmatrix} \sim \begin{bmatrix}1 & 2.5 & 4\\0 & 1 & 1\\0 & 0 & 0\\0 & 0 & 0\end{bmatrix} \sim \begin{bmatrix}1 & 0 & 1.5\\0 & 1 & 1\\0 & 0 & 0\\0 & 0 & 0\end{bmatrix}.$$

The desired nutrients are provided by 1.5 servings of Cheerios together with 1 serving of 100% Natural Cereal.

Study Tip: Be sure to distinguish between (i) the vector equation, (ii) the matrix equation (which has the form $A\mathbf{x} = \mathbf{b}$), and (iii) the augmented matrix (which has the form $[A\ \ \mathbf{b}]$) that represents a system of linear equations.

7. a. The fraction of the California population that left the state during 1990 was (509,500)/(29,716,000) = .01715 (shown to 5 decimal places in order to get four significant figures). The fraction that remained inside the state is 1 - .01715 = .98285.

The fraction of the population *outside* California that moved to California in 1990 was (564,100)/(218,994,000) = .002576 (stated to 6 decimal places in order to get four significant figures). The fraction that remained outside California is 1 - .002576 = .997424. So the migration matrix is

$$\begin{array}{c} \text{From:} \\ \begin{array}{cc} \text{Calif.} & \text{Outside} \end{array} \end{array}$$

$$M = \begin{bmatrix} .98285 & .002576 \\ .01715 & .997424 \end{bmatrix} \begin{array}{l} \text{To:} \\ \text{Calif.} \\ \text{Outside} \end{array}$$

b. $\mathbf{x}_{10} = \begin{bmatrix} 30,215,000 \\ 218,495,000 \end{bmatrix}$ To the nearest thousand

c. The total population of the United States in 1990 was 248,710,000. That number is also the total of the entries in $\mathbf{x}_{10}$ because the model does not consider growth of the population. Divide the entries in $\mathbf{x}_{10}$ by this number to get the percentage distribution of the U. S. population in the year 2000, namely, 12.15% in California and 87.85% elsewhere. At least that is the prediction of the model. Shortly after I obtained the data for this model, the population movement patterns changed, which means that the migration matrix entries are no longer correct. For the first time in recent history, the movement *out* of California exceeds the movement into the state. Whether that trend will change is uncertain.

MATLAB The M-file for Exercises 5-9 in Section 2.7 stores initial vectors in **x0**. Set **x = x0** to put the initial data into **x**. Then use the command **x = M*x** repeatedly to generate the sequence $\mathbf{x}_1, \mathbf{x}_2, \ldots$. You only type the command once. After that, use the up-arrow (↑) key to recall the command, and press <Enter>.

In Exercise 7, you need 6 decimal places to get four significant figures in M(1,2). Use the command **format long** and then **M** to see more decimal places in *M*. The command **format short** or simply **format** will return MATLAB to the standard display.

Numbers are entered in MATLAB without commas. The number 600,000 in MATLAB scientific notation is 6e5. A small number such as .00000012 is 1.2e-7.

CHAPTER 2 GLOSSARY CHECKLIST

Check your knowledge by attempting to write definitions of the terms below. Then compare your work with the definitions given in the text's Glossary. Ask your instructor which definitions, if any, might appear on a test.

affine transformation: A mapping $T:\mathbb{R}^n \to \mathbb{R}^m$ of the form $T(\mathbf{x}) = \ldots$.

collinear (vectors): Two or more vectors (points) that ....

contraction: A mapping $\mathbf{x} \mapsto \ldots$.

difference equation (or **linear recurrence relation**): An equation of the form ... whose solution is ....

dilation: A mapping $\mathbf{x} \mapsto \ldots$.

domain (of a transformation T): The set of ....

equal vectors: Vectors in $\mathbb{R}^n$ whose ....

Givens rotation: A linear transformation from $\mathbb{R}^n$ to $\mathbb{R}^n$ used in computer programs to ....

homogeneous equation: An equation of the form ....

identity matrix (denoted by I or I_n): A square matrix with ....

image (of a vector $\mathbf{x}$ under a transformation T): The vector ....

linear combination: A sum of ....

linear dependence relation: A ... equation where ....

linearly dependent (vectors): A set $\{\mathbf{v}_1, \ldots, \mathbf{v}_p\}$ with the property that ....

linearly independent (vectors): A set $\{\mathbf{v}_1, \ldots, \mathbf{v}_p\}$ with the property ....

linear transformation: A transformation $T:\mathbb{R}^n \to \mathbb{R}^m$ for which (i) ..., and (ii) ....

line through p parallel to v: The set ....

matrix equation: An equation that ....

matrix transformation: A mapping $\mathbf{x} \mapsto \ldots$.

migration matrix: A matrix that gives the ... movement between different locations, from ....

nontrivial solution: A ... solution of ....

one-to-one (mapping): A mapping $T:\mathbb{R}^n \to \mathbb{R}^m$ such that ....

onto (mapping): A mapping $T:\mathbb{R}^n \to \mathbb{R}^m$ such that ....

parallelogram rule for addition: A geometric interpretation of ....

parametric equation of a line: An equation of the form ....

parametric equation of a plane: An equation of the form ....

product $A\mathbf{x}$: ....

range (of a linear transformation T): The set of ....

rule for computing $A\mathbf{x}$: ....

scalar: ....

scalar multiple of $\mathbf{u}$ by c: The vector ....

set spanned by $\{\mathbf{v}_1, \ldots, \mathbf{v}_p\}$: ....

Span $\{\mathbf{v}_1, \ldots, \mathbf{v}_p\}$: The set ....

standard matrix (for a linear transformation T): The matrix ....

transformation (or **function** or **mapping**) T from $\mathbb{R}^n$ to $\mathbb{R}^m$: A rule that assigns to each vector $\mathbf{x}$ in $\mathbb{R}^n$ a .... Notation: $T: \mathbb{R}^n \longrightarrow \mathbb{R}^m$.

translation (by a vector $\mathbf{p}$): The operation of ....

trivial solution: The solution ... of a ....

vector: ....

vector equation: An equation involving ....

weights: ....

3

Matrix Algebra

3.1 MATRIX OPERATIONS

Most of this chapter is an outgrowth of the idea in Section 2.5 that a matrix can transform data. This dynamic role of matrices suggests that we study the *combined effect* of several matrices on data (that is, on a vector or a set of vectors). Sections 3.1 to 3.5 describe this *matrix algebra.*

KEY IDEA

Matrix multiplication corresponds to composition of linear transformations. The definition of AB, using the columns of B, is critical for the development of both the theory and some of the applications in the text.

STUDY NOTES

Double-subscript notation: The subscripts tell the location of an entry in the matrix—the first subscript identifies the row and the second subscript the column. (Remember: *Row* is shorter than *column*, so *row* goes first.)

In the product AB, left multiplication (that is, multiplication on the left) by A acts on the columns of B, by definition, while right multiplication by B acts on the rows of A (see page 102). That is,

$$\begin{bmatrix}\text{column } j\\ \text{of } AB\end{bmatrix} = A\begin{bmatrix}\text{column } j\\ \text{of } B\end{bmatrix} \quad \text{and} \quad [\text{row } i \text{ of } AB] = [\text{row } i \text{ of } A]B$$

To compute a specific matrix product by hand, use the Row-Column Rule. If A is $m\times n$, then the (i,j)-entry of AB is written with sigma notation as

$$(AB)_{ij} = \sum_{k=1}^{n} a_{ik}b_{kj}$$

Remember that if you change the *order* (position) of the factors in a matrix product, the new product may be different, or it may not even be defined. For instance, $(A + C)B$ and $AB + BC$ are probably *not* equal! Also, see the warning box on page 104.

SOLUTIONS TO EXERCISES

1. $-2A = (-2)\begin{bmatrix} 7 & 0 & -1 \\ -1 & 5 & 2 \end{bmatrix} = \begin{bmatrix} -14 & 0 & 2 \\ 2 & -10 & -4 \end{bmatrix}$

7. $CD = \begin{bmatrix} 1 & 4 \\ -4 & 0 \end{bmatrix}\begin{bmatrix} 1 & 0 \\ -2 & 1 \end{bmatrix} = \begin{bmatrix} -7 & 4 \\ -4 & 0 \end{bmatrix}$

13. Since A has 5 columns, B must have 5 rows. Otherwise, AB is not defined. Since AB has 7 columns, so does B. Thus B is 5×7.

19. $AD = \begin{bmatrix} 1 & 1 & 1 \\ 1 & 2 & 3 \\ 1 & 4 & 5 \end{bmatrix}\begin{bmatrix} 2 & 0 & 0 \\ 0 & 3 & 0 \\ 0 & 0 & 4 \end{bmatrix} = \begin{bmatrix} 2 & 3 & 4 \\ 2 & 6 & 12 \\ 2 & 12 & 20 \end{bmatrix}$

$DA = \begin{bmatrix} 2 & 0 & 0 \\ 0 & 3 & 0 \\ 0 & 0 & 4 \end{bmatrix}\begin{bmatrix} 1 & 1 & 1 \\ 1 & 2 & 3 \\ 1 & 4 & 5 \end{bmatrix} = \begin{bmatrix} 2 & 2 & 2 \\ 3 & 6 & 9 \\ 4 & 16 & 20 \end{bmatrix}$

Right-multiplication by D multiplies the columns in A by the corresponding diagonal entries in D. Left-multiplication by D multiplies the *rows* in A by the corresponding diagonal entries in D.

If λ (Greek: *lambda*) is any scalar, then $(\lambda I)A = \lambda A = A(\lambda I)$. (Multiplying *all* the rows of A by λ is the same as multiplying *all* the columns by λ.)

25. Let $\mathbf{b}_p$ be the last column of B. Then $A\mathbf{b}_p = \mathbf{0}$, but $\mathbf{b}_p$ is not the zero vector. Thus the equation $A\mathbf{b}_p = \mathbf{0}$ is a linear dependence relation, and the columns of A are linearly dependent.

Supplementary Exercises: The following two problems will help you review concepts from Chapter 2. Try them now. The answers are at the end of this section. In each problem, assume that the product AB is defined.

a. Show that if $\mathbf{y}$ is a linear combination of the columns of AB, then $\mathbf{y}$ is a linear combination of the columns of A.

b. Show that if the columns of B are linearly dependent, then so are the columns of AB.

31. $\mathbf{u}^T\mathbf{v} = [-2 \quad 3 \quad -4]\begin{bmatrix} a \\ b \\ c \end{bmatrix} = -2a + 3b - 4c$ (a 1×1 matrix)

$$\mathbf{v}^T\mathbf{u} = [a \quad b \quad c]\begin{bmatrix} -2 \\ 3 \\ -4 \end{bmatrix} = a(-2) + b(3) + c(-4) = -2a + 3b - 4c = \mathbf{u}^T\mathbf{v}$$

$$\mathbf{u}\mathbf{v}^T = \begin{bmatrix} -2 \\ 3 \\ -4 \end{bmatrix}[a \quad b \quad c] = \begin{bmatrix} -2a & -2b & -2c \\ 3a & 3b & 3c \\ -4a & -4b & -4c \end{bmatrix} \quad \text{Outer product}$$

$$\mathbf{v}\mathbf{u}^T = \begin{bmatrix} a \\ b \\ c \end{bmatrix}[-2 \quad 3 \quad -4] = \begin{bmatrix} a(-2) & a(3) & a(-4) \\ b(-2) & b(3) & b(-4) \\ c(-2) & c(3) & c(-4) \end{bmatrix} = \begin{bmatrix} -2a & 3a & -4a \\ -2b & 3b & -4b \\ -2c & 3c & -4c \end{bmatrix} = (\mathbf{u}\mathbf{v}^T)^T$$

Study Tip: *Inner* products ($\mathbf{u}^T\mathbf{v}$ and $\mathbf{v}^T\mathbf{u}$) have the transpose symbol in the middle. *Outer* products ($\mathbf{u}\mathbf{v}^T$ and $\mathbf{v}\mathbf{u}^T$) have the transpose symbol on the outside.

35. Let r be a scalar, and denote the jth columns of A and B by $\mathbf{a}_j$ and $\mathbf{b}_j$, respectively. Then the jth column of $r(A + B)$ is $r(\mathbf{a}_j + \mathbf{b}_j) = r\mathbf{a}_j + r\mathbf{b}_j$. Since $r\mathbf{a}_j$ is the jth column of rA, and $r\mathbf{b}_j$ is the jth column of rB, we have shown that the jth columns of $r(A + B)$ and $rA + rB$ are equal. This is true for each column, so Theorem 1(d) is proved.

37. The (i, j)-entry of $A(B + C)$ equals the (i, j)-entry of $AB + AC$, because

$$\sum_{k=1}^{n} a_{ik}(b_{kj} + c_{kj}) = \sum_{k=1}^{n} a_{ik}b_{kj} + \sum_{k=1}^{n} a_{ik}c_{kj}$$

39. The (i, j)-entries of $r(AB)$, $(rA)B$, and $A(rB)$ are all equal, because

$$r\sum_{k=1}^{n} a_{ik}b_{kj} = \sum_{k=1}^{n} (ra_{ik})b_{kj} = \sum_{k=1}^{n} a_{ik}(rb_{kj})$$

41. Let $\mathbf{e}_j$ and $\mathbf{a}_j$ denote the jth columns of I_n and A, respectively. Since $\mathbf{e}_j$ has 1 in the jth position and zeros elsewhere, $A\mathbf{e}_j = \mathbf{a}_j$, by definition of matrix-vector multiplication. So

$$AI_n = A[\mathbf{e}_1 \cdots \mathbf{e}_n] = [A\mathbf{e}_1 \cdots A\mathbf{e}_n] = [\mathbf{a}_1 \cdots \mathbf{a}_n] = A$$

Study Tip: Make a note to remember why $A\mathbf{e}_j = \mathbf{a}_j$. You may need this later.

43. The hint in the text leads to a rather inefficient proof. Sorry. Here is a better proof: The (r,s)-entry in $(AB)^T$ is the (s,r)-entry in AB, which is

$$a_{s1}b_{1r} + \cdots + a_{sn}b_{nr}$$

The entries in row r of B^T are $b_{1r}, \cdots, b_{nr}$, because they come from the rth column of B. Likewise, the entries in column s of A^T are $a_{s1}, \ldots, a_{sn}$, because they come from the sth row of A. Thus the (r,s)-entry in B^TA^T is $b_{1r}a_{s1} + \cdots + b_{nr}a_{sn}$. This sum equals the other sum displayed, so the (r,s)-entries of $(AB)^T$ and B^TA^T are equal. This proves Theorem 3(d).

Solutions to Supplementary Exercises:

a. If $\mathbf{y}$ is a linear combination of the columns of AB, then the equation $AB\mathbf{x} = \mathbf{y}$ has a solution. (See the box on page 50 in the text.) If $\mathbf{x}$ is such a solution, then associativity of matrix multiplication shows that $A(B\mathbf{x}) = \mathbf{y}$. That is, the equation $A\mathbf{u} = \mathbf{y}$ has a solution. So $\mathbf{y}$ is a linear combination of the columns of A.

b. If the columns of B are linearly dependent, then there is a nonzero $\mathbf{x}$ such that $B\mathbf{x} = \mathbf{0}$. Hence $A(B\mathbf{x}) = A\mathbf{0} = \mathbf{0}$, and $(AB)\mathbf{x} = \mathbf{0}$, by associativity, which shows that the columns of AB are linearly dependent, because $\mathbf{x}$ is not zero.

MATLAB If A is $m \times n$, then **size(A)** is the row vector $[m \;\; n]$. The (i,j)-entry in A is **A(i,j)**. If i or j is replaced by a colon, the result is a column or row of A. Examples:

A(:,3) is column 3 of A
A(2,:) is row 2 of A

To create a matrix, enter the data row-by-row, with a space between entries and a semicolon between rows. For instance, the command

A = [1 2 3;4 5 -6] Use brackets around the data.

creates a 2×3 matrix A.

MATLAB uses +, -, and * to denote matrix addition, subtraction, and multiplication, respectively. If k is a positive integer, **A^k** denotes the kth power of A. The transpose of A is **A'** (using an apostrophe for the prime symbol). Vectors are treated as (single) row or column matrices. If **u** and **v** are column vectors of the same size, then **u'*v** is their inner product, and **u*v'** is an outer product. Note: the prime symbol creates the transpose when a vector or matrix has only real entries. For complex vectors or matrices, see a MATLAB manual.

3.2 THE INVERSE OF A MATRIX

Matrix inverses are essential for many discussions in linear algebra. This section and the next describe the main properties of invertible matrices.

KEY IDEAS

The inverse formula for a 2×2 matrix will be used frequently in exercises later in the text. To invert a 2×2 matrix, interchange the diagonal entries, reverse the signs of the off-diagonal entries, and divide each entry by the determinant (assuming $ad - bc \neq 0$).

Theorem 5 and its proof are important. The phrase "has a unique solution" includes the assertion that a solution exists, so the proof has two parts. The equation $AA^{-1} = I$ is used to prove that a solution exists, and the equation $A^{-1}A = I$ is used to show that the solution is unique.

Except when A is 2×2, Theorem 5 is practically never used to solve $A\mathbf{x} = \mathbf{b}$. Row reduction of $[A \;\; \mathbf{b}]$ is faster. Actually, in practical work, you will seldom need to compute A^{-1}. (However, Example 3 illustrates a case in which the entries of A^{-1} could be useful.)

When using an inverse in matrix algebra, remember that matrix multiplication is not commutative. The phrase "left-multiply B by A^{-1}" means to multiply B on its left side by A^{-1}. *Never* write $\frac{B}{A}$ (or B/A) because it could stand for $A^{-1}B$ or BA^{-1}.

Elementary matrices are used in this text mainly to link row reduction to matrix multiplication. Each elementary row operation amounts to left-multiplication by an elementary matrix.

SOLUTIONS TO EXERCISES

1. $$\begin{bmatrix} 3 & -8 \\ -1 & 3 \end{bmatrix}^{-1} = \frac{1}{3\cdot 3 - (-8)(-1)}\begin{bmatrix} 3 & -(-8) \\ -(-1) & 3 \end{bmatrix} = \begin{bmatrix} 3 & 8 \\ 1 & 3 \end{bmatrix}$$

7. a. $$A^{-1} = \frac{1}{1\cdot 8 - 2\cdot 3}\begin{bmatrix} 8 & -2 \\ -3 & 1 \end{bmatrix} = \frac{1}{2}\begin{bmatrix} 8 & -2 \\ -3 & 1 \end{bmatrix} \text{ or } \begin{bmatrix} 4 & -1 \\ -3/2 & 1/2 \end{bmatrix}$$

$$\mathbf{x} = A^{-1}\mathbf{b}_1 = \frac{1}{2}\begin{bmatrix} 8 & -2 \\ -3 & 1 \end{bmatrix}\begin{bmatrix} 5 \\ 7 \end{bmatrix} = \frac{1}{2}\begin{bmatrix} 26 \\ -8 \end{bmatrix} = \begin{bmatrix} 13 \\ -4 \end{bmatrix}$$

Similar calculations give $A^{-1}\mathbf{b}_2 = \begin{bmatrix} -6 \\ 2 \end{bmatrix}$, $A^{-1}\mathbf{b}_3 = \begin{bmatrix} 7 \\ -2 \end{bmatrix}$, $A^{-1}\mathbf{b}_4 = \begin{bmatrix} -3 \\ 2 \end{bmatrix}$.

b. $[A \quad \mathbf{b}_1 \quad \mathbf{b}_2 \quad \mathbf{b}_3 \quad \mathbf{b}_4] = \begin{bmatrix} 1 & 2 & 5 & -2 & 3 & 1 \\ 3 & 8 & 7 & -2 & 5 & 7 \end{bmatrix}$

$$\sim \begin{bmatrix} 1 & 2 & 5 & -2 & 3 & 1 \\ 0 & 2 & -8 & 4 & -4 & 4 \end{bmatrix} \sim \begin{bmatrix} 1 & 2 & 5 & -2 & 3 & 1 \\ 0 & 1 & -4 & 2 & -2 & 2 \end{bmatrix}$$

$$\sim \begin{bmatrix} 1 & 0 & 13 & -6 & 7 & -3 \\ 0 & 1 & -4 & 2 & -2 & 2 \end{bmatrix}$$

The solutions are $\begin{bmatrix} 13 \\ -4 \end{bmatrix}, \begin{bmatrix} -6 \\ 2 \end{bmatrix}, \begin{bmatrix} 7 \\ -2 \end{bmatrix}, \begin{bmatrix} -3 \\ 2 \end{bmatrix}$, the same as in part (a).

Note: This exercise was designed to make the arithmetic simple for both methods, but (a) requires more arithmetic than (b). In fact, (a) requires 22 multiplications or divisions and 9 additions or subtractions, but (b) only takes 14 multiplications or divisions and 9 additions or subtractions.

In general, the arithmetic for method (b) can be unpleasant for hand calculation. However, when A is larger than 2×2, method (b) is *much* faster than (a).

Study Tip: Notice in Exercise 7(a) how the 1/2 in the formula for A^{-1} was kept outside the matrix $\begin{bmatrix} 8 & -2 \\ -3 & 1 \end{bmatrix}$ when computing $A^{-1}\mathbf{x}$. This trick often simplifies hand calculation by postponing the arithmetic with fractions (or decimals) until the end.

9. a. Given: A is $n \times n$ and invertible, and B is $n \times p$. Suppose that X_1 satisfies $AX_1 = B$. Then, left-multiplying each side by A^{-1},

$$A^{-1}(AX_1) = A^{-1}B, \quad IX_1 = A^{-1}B, \quad \text{and} \quad X_1 = A^{-1}B$$

If there is a solution of $AX = B$, then the solution must be $A^{-1}B$. To show that $A^{-1}B$ *is* a solution, substitute it for X and compute

$$A(A^{-1}B) = AA^{-1}B = IB = B$$

b. Write $B = [\mathbf{b}_1 \ \cdots \ \mathbf{b}_p]$ and $X = [\mathbf{u}_1 \ \cdots \ \mathbf{u}_p]$. By definition of matrix multiplication, $AX = [A\mathbf{u}_1 \ \cdots \ A\mathbf{u}_p]$. So the equation $AX = B$ is equivalent to the p systems:

$$A\mathbf{u}_1 = \mathbf{b}_1, \quad \ldots \quad , \quad A\mathbf{u}_p = \mathbf{b}_p \tag{1}$$

Since A is the coefficient matrix in each system, these systems may be solved simultaneously, placing the augmented columns of these systems next to A to form $[A \quad \mathbf{b}_1 \quad \cdots \quad \mathbf{b}_p] = [A \quad B]$. Since the solutions $\mathbf{u}_1, \ldots, \mathbf{u}_p$ in (1) are uniquely determined, we know that $[A \quad \mathbf{b}_1 \quad \cdots \quad \mathbf{b}_p]$ reduces to $[I \quad \mathbf{u}_1 \quad \cdots \quad \mathbf{u}_p]$, which is $[I \quad X]$, where X is the solution of $AX = B$.

Another solution: Since A is invertible, it can be row reduced to I. Thus $[A \quad B]$ can be row reduced to a matrix of the form $[I \quad X]$ for some X. Let $E_1, \ldots, E_k$ be elementary matrices that implement this row reduction. Then

$$(E_k \cdots E_1)A = I \quad \text{and} \quad (E_k \cdots E_1)B = X \tag{2}$$

Right-multiplying each side of the first equation by A^{-1}, we find that $(E_k \cdots E_1)AA^{-1} = IA^{-1}$, $(E_k \cdots E_1)I = A^{-1}$, and $E_k \cdots E_1 = A^{-1}$. The second equation in (2) then shows that $A^{-1}B = X$.

Study Tip: Whenever you are told "A is invertible", you know that A^{-1} exists, and you may use A^{-1} to solve an equation or to make appropriate calculations.

13. $AB = AC \;\Rightarrow\; A^{-1}AB = A^{-1}AC$ Because A is invertible

$\Rightarrow\; IB = IC \;\Rightarrow\; B = C$

Warning: A common mistake in Exercise 16 is to try to use the formula $(AB)^{-1} = B^{-1}A^{-1}$. But this formula is available only when you know, in advance, that A and B are invertible. In Exercise 16, you must *prove* that A is invertible.

19. $C^{-1}(A + X)B^{-1} = I \;\Rightarrow\; CC^{-1}(A + X)B^{-1} = CI$ Left-multiplication by C

$\Rightarrow I(A + X)B^{-1} = C \;\Rightarrow\; (A + X)B^{-1}B = CB$ Right-multiplication by B

$\Rightarrow\; (A + X)I = CB \;\Rightarrow\; A + X = CB \;\Rightarrow\; X = CB - A$

(Be careful above when you multiply by B. The right side must be CB, not BC.) To shows that $CB - A$ *is* a solution, substitute it for X:

$$C^{-1}[A + (CB - A)]B^{-1} = C^{-1}[CB]B^{-1} = C^{-1}CBB^{-1} = II = I$$

After this section, your instructor may permit you to include fewer details in your calculations (check on this). For instance, the argument above could be shortened to:

$C^{-1}(A + X)B^{-1} = I \;\Leftrightarrow\; (A + X)B^{-1} = C$ Left-multiplication by C or C^{-1}

$\Leftrightarrow\; (A + X) = CB$ Right-multiplication by B or B^{-1}

$\Leftrightarrow\; X = CB - A$

21. If $A\mathbf{x} = \mathbf{0}$ has more than one solution, then A cannot be invertible, by Theorem 5. If a, b, c, d are all zero, then A is the zero matrix and the equation $A\mathbf{x} = \mathbf{0}$ is true for *all* $\mathbf{x}$. Otherwise, at least one of the vectors $\begin{bmatrix} -b \\ a \end{bmatrix}$ and $\begin{bmatrix} -d \\ c \end{bmatrix}$ is nonzero, and both of them satisfy $A\mathbf{x} = \mathbf{0}$. For example, $A\mathbf{x} = \begin{bmatrix} a & b \\ c & d \end{bmatrix}\begin{bmatrix} -b \\ a \end{bmatrix} = \begin{bmatrix} -ab + ba \\ -cb + da \end{bmatrix} = \begin{bmatrix} 0 \\ 0 \end{bmatrix}$, because $ad - bc = 0$.

25. $$EA = \begin{bmatrix} 1 & 0 & 0 & 0 \\ 0 & 1 & 0 & 0 \\ 0 & 0 & 5 & 0 \\ 0 & 0 & 0 & 1 \end{bmatrix}\begin{bmatrix} 1 & 0 & 0 & 0 \\ 2 & 1 & 0 & 0 \\ 3 & 2 & 1 & 0 \\ 4 & 3 & 2 & 1 \end{bmatrix} = \begin{bmatrix} 1 & 0 & 0 & 0 \\ 2 & 1 & 0 & 0 \\ 15 & 10 & 5 & 0 \\ 4 & 3 & 2 & 1 \end{bmatrix}$$

This result is also produced by multiplying the third row of A by 5. Left-multiplication by E would have the same effect on any 4×4 matrix. Note that this result also follows from Exercise 19 in Section 3.1, because E is a diagonal matrix: each row of A is multiplied by the corresponding diagonal entry of E.

31. $$\begin{bmatrix} 1 & 0 & 5 & 1 & 0 & 0 \\ 1 & 1 & 0 & 0 & 1 & 0 \\ 3 & 2 & 6 & 0 & 0 & 1 \end{bmatrix} \sim \begin{bmatrix} 1 & 0 & 5 & 1 & 0 & 0 \\ 0 & 1 & -5 & -1 & 1 & 0 \\ 0 & 2 & -9 & -3 & 0 & 1 \end{bmatrix} \sim$$

$$\begin{bmatrix} 1 & 0 & 5 & 1 & 0 & 0 \\ 0 & 1 & -5 & -1 & 1 & 0 \\ 0 & 0 & 1 & -1 & -2 & 1 \end{bmatrix} \sim \begin{bmatrix} 1 & 0 & 0 & 6 & 10 & -5 \\ 0 & 1 & 0 & -6 & -9 & 5 \\ 0 & 0 & 1 & -1 & -2 & 1 \end{bmatrix}. \quad A^{-1} = \begin{bmatrix} 6 & 10 & -5 \\ -6 & -9 & 5 \\ -1 & -2 & 1 \end{bmatrix}$$

37. The deflection vector is $\begin{bmatrix} .005 & .002 & .001 \\ .002 & .004 & .002 \\ .001 & .002 & .005 \end{bmatrix}\begin{bmatrix} 20 \\ 50 \\ 10 \end{bmatrix} = \begin{bmatrix} .21 \\ .26 \\ .17 \end{bmatrix}$. The deflections are .21, .26 and .17 inches.

Study Tip: In Exercise 38, the arithmetic to invert D is not bad by hand. It may help to multiply D by 1000 first. Then row operations will produce the inverse of $1000D$. But $(1000D)^{-1} = 1000^{-1}D^{-1}$. So multiply the inverse matrix you find by 1000 to obtain D^{-1}. Of course, using MATLAB is easier.

MATLAB The $n \times n$ identity matrix is denoted by eye(n). If A is $n \times n$, then the command **M = [A eye(n)]** creates the augmented matrix $[A \;\; I]$.
To speed up row reduction of $M = [A \;\; I]$, the command **gauss(M,p)** will use the leading entry in row p of M as a pivot, and use row replacements to create zeros in the pivot column below this pivot entry. The result is stored in the default matrix "ans", unless you assign the result to some other variable, such as M itself. For the backward

phase of row reduction, use the command **gauss(M,p,v)**, where *p* is the pivot row, and **v** is a vector (row or column) that lists the indices of the rows that need to have zeros in the pivot column. For instance, the command **M = gauss(M,5,[1 2 3 4])** will create zeros in rows 1 to 4 of the pivot column (above the pivot in row 5). Another command that does the same thing is **M = gauss(M,5,1:4)**, because the symbols 1:4 (read "1 to 4") stand for the vector [1 2 3 4]. The command **gauss** is on the Toolbox data diskette.

There are other MATLAB commands that row reduce matrices, invert matrices, and solve equations $A\mathbf{x} = \mathbf{b}$, but they are not discussed here because they will not help you learn the concepts in this section.

3.3 CHARACTERIZATIONS OF INVERTIBLE MATRICES

Now is the time to review all the major concepts presented in previous sections. Compared to other sections, this one is half as long and needs more than twice as much study time. **Don't postpone your work on this material.** You must master the ideas before your reach Chapter 5. (In many texts, the equivalent of Chapter 5 is usually the "killer" chapter. But you won't have difficulties with your text if you prepare well now.)

KEY IDEAS

The Invertible Matrix Theorem (IMT) only applies to square matrices. However, some groups of these statements in the IMT are also equivalent for rectangular matrices. See Theorem 2 in Section 2.2, for example.

STATEMENTS FROM THE INVERTIBLE MATRIX THEOREM

Equivalent statements, for any $n \times p$ matrix A.	Equivalent only for an $n \times n$ square matrix A.	Equivalent statements, for any $n \times p$ matrix A.
f. There is a matrix C such that $AC = I$.	a. A is an invertible matrix.	g. There is a matrix D such that $DA = I$.
*. A has a pivot position in every row.	b′. A has n pivot positions.	*. A has a pivot position in every column.
e″. The columns of A span $\mathbb{R}^n$.	b. A is row equivalent to the $n \times n$ identity matrix.	d′. The columns of A form a linearly independent set.

*. The equation $A\mathbf{x} = \mathbf{b}$ has at least one solution for each $\mathbf{b}$ in $\mathbb{R}^n$.	e′. The equation $A\mathbf{x} = \mathbf{b}$ has a unique solution for each $\mathbf{b}$ in $\mathbb{R}^n$.	*. The equation $A\mathbf{x} = \mathbf{b}$ has at most one solution for each $\mathbf{b}$ in $\mathbb{R}^n$.
e‴. The linear transformation $\mathbf{x} \mapsto A\mathbf{x}$ maps $\mathbb{R}^p$ onto $\mathbb{R}^n$.	*. The linear transformation $\mathbf{x} \mapsto A\mathbf{x}$ is invertible.	d". The linear transformation $\mathbf{x} \mapsto A\mathbf{x}$ is one-to-one.
	c. A is a product of elementary matrices.	d. The equation $A\mathbf{x} = \mathbf{0}$ has only the trivial solution.
	h. A^T is an invertible matrix.	*. The equation $A\mathbf{x} = \mathbf{0}$ has no free variables.

The six statements denoted by (*) were not listed in the text as part of the IMT, mainly to avoid intimidating you with so many statements in one theorem. As part of your review, make sure you understand why these extra statements can be added to the IMT.

The text did not actually prove that for a *rectangular* matrix, statements (f) and (g) are equivalent to the other statements in their respective columns, but I listed them here anyway so you could see how they fit into the table. A matrix D such that $DA = I$ is called a **left-inverse** of A, and a matrix C such that $AC = I$ is called a **right-inverse** of A.

Checkpoint: What can you say about the statements in the first column when A has more rows than columns? (Why?) What about the statements in the third column when A has more columns than rows? (Why?)

A question such as the one in the box below is one way I test whether my students know the IMT. Test yourself. Cover up the IMT, write your answer, and then check your work. The answer is given at the end of this section.

Test Question:

Write 6 statements from the Invertible Matrix Theorem, each equivalent to the statement that an $n \times n$ matrix A is invertible. Use the following concepts, one in each statement: (*i*) row equivalent, (*ii*) the equaion $AC = I$, (*iii*) columns, (*iv*) the equation $A\mathbf{x} = \mathbf{0}$, (*v*) elementary matrices, and (*vi*) linear transformation.

SOLUTIONS TO EXERCISES

1. To show that A is not invertible, you have to find only **one** statement in the Invertible Matrix Theorem that is not true. For instance, the text's answer used the obvious fact that the columns of A are not linearly independent.

The text also showed how to use $\det A$ and Theorem 4 to show that A is not invertible. Until Chapter 4, this approach is available only for 2×2 matrices.

7. $\begin{bmatrix} 5 & -9 & 3 \\ 0 & 3 & 4 \\ 1 & 0 & 3 \end{bmatrix} \sim \begin{bmatrix} 1 & 0 & 3 \\ 0 & 3 & 4 \\ 5 & -9 & 3 \end{bmatrix} \sim \begin{bmatrix} 1 & 0 & 3 \\ 0 & 3 & 4 \\ 0 & -9 & -12 \end{bmatrix} \sim \begin{bmatrix} 1 & 0 & 3 \\ 0 & 3 & 4 \\ 0 & 0 & 0 \end{bmatrix}$. The 3×3 matrix has only 2 pivot positions and thus is not invertible, by the IMT. (These calculations also show that the matrix cannot be row equivalent to I, as mentioned in the text.)

Study Tip: For theoretical purposes, the most useful characterization of an invertible matrix A is that $A\mathbf{x} = \mathbf{0}$ has only the trivial solution. For most computational work by hand, the fastest way to determine if A is invertible is to (find and) count the pivot positions.

13. A square upper triangular matrix is invertible if and only if the entries on the diagonal are nonzero. The only way an $n\times n$ upper triangular matrix can have n pivots is to have nonzero entries on its diagonal.

If A is an $n\times n$ lower triangular matrix, then A^T is upper triangular, with the same diagonal entries. Thus A^T is invertible if and only if the diagonal entries of A are nonzero. But A^T is invertible if and only if A is invertible (by the IMT), so the diagonal entries of A are nonzero if and only if A is invertible.

19. The full solution is in the text's answer section.

Study Tip: Learn how to recognize when a square matrix is **not** invertible. See Exercises 15 and 19. Each of the following statements was constructed by *negating* one of the statements in the IMT. If A is an $n\times n$ matrix, then each statment is true if and only if A is **not** invertible:

- The matrix A has *fewer than* n pivot positions.
- The equation $A\mathbf{x} = \mathbf{0}$ has a *nontrivial* (nonzero) solution.
- The equation $A\mathbf{x} = \mathbf{b}$ has *more than one* solution *for some* $\mathbf{b}$ in $\mathbb{R}^n$.
- The equation $A\mathbf{x} = \mathbf{b}$ has *no* solution (is inconsistent) *for some* $\mathbf{b}$ in $\mathbb{R}^n$.
- The columns of A are linearly *dependent*.
- The columns of A *do not* span $\mathbb{R}^n$.

21. By Theorem 6(b) in Section 3.2 (with A in place of B), the product AA is invertible, because both A and A are invertible.

23. Suppose that A is square and $AB = I$. Then A is invertible, by the IMT. Left-multiplying the equation $AB = I$ by A^{-1}, we have $A^{-1}AB = A^{-1}I$, $IB = A^{-1}$, and $B = A^{-1}$. Taking inverses, $B^{-1} = (A^{-1})^{-1} = A$.

Next, suppose $BA = I$. You could mimic the argument just given, using the IMT and then right-multiplying by A^{-1} to get $B = A^{-1}$. Another argument: The equation $BA = I$ is the same as the one just studied except that the roles of A and B are interchanged. So, by what we have already proved, the conclusion holds with A and B interchanged, namely, "B and A are invertible, with $A = B^{-1}$ and $B = A^{-1}$." This is the conclusion we need! (This argument is rather sophisticated, but it is often used in mathematics when some statement is not affected by the interchange of two symbols.)

Study Tip: Exercise 23 makes a good test qustion.

25. If AB is invertible, define $C = B(AB)^{-1}$. Then

$$AC = AB(AB)^{-1} = I \qquad (*)$$

To show that A is invertible, you could try to compute the product $CA = B(AB)^{-1}A$. However, this product cannot be simplified unless you already know that C is A^{-1}. The way out of this predicament is to observe that since A is square, equation (*) and the Invertible Matrix Theorem show that A is invertible.

Another proof: If AB is invertible, then there is a matrix M such that $(AB)M = I$ and hence $A(BM) = I$. Since A is square, A is invertible, by the IMT. (My friend, Prof. Richard Good, showed me this nice proof after the text was published!)

Study Tip: The solution to Exercise 25 shows a common **use** of the IMT. Just remember that statements (f) and (g) of the IMT apply only to square matrices. Whenever you use one of these statements to deduce that A is invertible, you should point out in your argument that A is square.

27. To show that T is one-to-one, suppose that $T(\mathbf{u}) = T(\mathbf{v})$ for some vectors $\mathbf{u}$ and $\mathbf{v}$ in $\mathbb{R}^n$. Then $S(T(\mathbf{u})) = S(T(\mathbf{v}))$, where S is the inverse of T. By Equation (1), $\mathbf{u} = S(T(\mathbf{u}))$ and $S(T(\mathbf{v})) = \mathbf{v}$, so $\mathbf{u} = \mathbf{v}$. Thus T is one-to-one. To show that T is onto, suppose $\mathbf{y}$ represents an arbitrary vector in $\mathbb{R}^n$ and define $\mathbf{x} = S\mathbf{y}$. Then, using Equation (2), $T(\mathbf{x}) = T(S(\mathbf{y})) = \mathbf{y}$, which shows that T maps $\mathbb{R}^n$ onto $\mathbb{R}^n$.

Second proof: By Theorem 9, the standard matrix A of T is invertible. By the IMT, the columns of A are linearly independent and span $\mathbb{R}^n$. Finally, by Theorem 10 in Section 2.6, T is one-to-one and maps $\mathbb{R}^n$ onto $\mathbb{R}^n$.

29. The standard matrix of T is $A = \begin{bmatrix} -5 & 9 \\ 4 & -7 \end{bmatrix}$, which is invertible, because $\det A = -1 \neq 0$. By Theorem 9, the transformation T is invertible, and its standard matrix is A^{-1}. From the formula for a 2×2 inverse, $A^{-1} = \begin{bmatrix} 7 & 9 \\ 4 & 5 \end{bmatrix}$. So $T^{-1}(x_1, x_2) = \begin{bmatrix} 7 & 9 \\ 4 & 5 \end{bmatrix}\begin{bmatrix} x_1 \\ x_2 \end{bmatrix} = (7x_1 + 9x_2, 4x_1 + 5x_2)$.

31. Given any $\mathbf{v}$ in $\mathbb{R}^n$, we may write $\mathbf{v} = T(\mathbf{x})$ for some $\mathbf{x}$, because T is an onto mapping. Then, the assumed properties of S and U show that $S(\mathbf{v}) = S(T(\mathbf{x})) = \mathbf{x}$ and $U(\mathbf{v}) = U(T(\mathbf{x})) = \mathbf{x}$. So $S(\mathbf{v})$ and $U(\mathbf{v})$ are equal for each $\mathbf{v}$. That is, S and U are the same function from $\mathbb{R}^n$ into $\mathbb{R}^n$.

Answers to Checkpoint: If A has more rows than columns, then all statements in the first column of the table must be false, because they are equivalent and the statement about a pivot position in each row cannot be true. If A has more columns than rows, then all statements in the third column of the table must be false, because A cannot have a pivot in each of its columns.

Answers to Test Question:

(*i*) A is row equivalent to I_n. (*ii*) There exists an $n \times n$ matrix C such that $AC = I$. (*iii*) The columns of A span $\mathbb{R}^n$. (*iv*) The equation $A\mathbf{x} = \mathbf{0}$ has only the trivial solution. (*v*) A is a product of elementary matrices. (*vi*) The linear transformation $\mathbf{x} \mapsto A\mathbf{x}$ is one-to-one.

Another answer for (*iii*) is: The columns of A are linearly independent. Similarly, (*vi*) has another answer. But the following statement is unacceptable as one of the answers to the test question:

A is invertible if and only if the columns of A span $\mathbb{R}^n$.

This statement is itself a (true) theorem (assuming A is square), not a statement that is true precisely when A is invertible.

MATLAB: Determining whether a specific numerical matrix is invertible is not always a simple matter. A fast and fairly reliable method is to use the command **inv(A)** , which computes the inverse of A. A warning is given if the matrix is "singular" (noninvertible) or close to singular.

Another method to test invertibility is to enter **rank(A)** . Section 5.6 discusses rank and shows that an $n \times n$ matrix A is invertible if and only if $\text{rank}(A) = n$. The MATLAB **rank** command uses the singular value decomposition (see Section 8.4), which requires more calculations than **inv** , but is based on a particularly accurate algorithm.

3.4 PARTITIONED MATRICES

The ideas in this section are fairly simple. However, mark them for future reference, because you are likely to use this notation after you leave school. Partitioned matrices arise in theoretical discussions in essentially every field that makes use of matrices. Here are two examples.

1. The modern *state space* approach to control systems engineering depends on matrix calculations.[1] The problem of determining whether a system is *controllable* amounts to calculating the number of pivot positions in a *controllability matrix*

$$[B \quad AB \quad A^2B \quad \cdots \quad A^{n-1}B]$$

   where A is $n\times n$, B has n rows, and the matrices come from an equation of the form (8) in the discussion preceding Exercise 17.

2. Discussions of modern algorithms and computer software design for scientific computing naturally use the "language" of partitioned matrices. For instance, common techniques for parallel processing of large matrix calculations, such as *slicing* and *crinkling*, are described with partitioned matrices.[2] Also, the standard computer science reference on matrix calculations relies heavily on partitioned matrices.[3]

KEY IDEAS

The column-row evaluation of AB is the last of five different "views" of matrix multiplication. All five are special cases of the *block matrix* version of the *row-column* rule for matrix multiplication. Here they are:

---

[1]An understanding of control systems is important in the design of filtering circuitry, robots, process control systems, and spacecraft. Thus a control systems course is often part of the undergraduate curriculum for electrical, mechanical, chemical, and aerospace engineering. See *Control Systems Engineering*, by Norman S. Nise, Benjamin/Cummings Publishing Co., Redwood City, CA, 1992.

[2]*Parallel Algorithms and Matrix Computations*, by Jagdish J. Modi, Oxford Applied Mathematics and Computing Science Series, Clarendon Press, Oxford, 1988, pp. 73-75.

[3]*Matrix Computations*, 2nd ed., by Gene H. Golub and Charles F. Van Loan, The Johns Hopkins Press, Baltimore, 1989.

(1) The definition of $A\mathbf{x}$ amounts to block multiplication of AB where B has only one column:

$$A\mathbf{x} = \left[\mathbf{a}_1 \mid \cdots \mid \mathbf{a}_n\right]\begin{bmatrix} x_1 \\ \hline \vdots \\ \hline x_n \end{bmatrix} = \left[x_1\mathbf{a}_1 + \cdots + x_n\mathbf{a}_n\right]$$

(2) Partition A as *one* row and *one* column. Then the definition of the usual product is a row-column block product:

$$AB = A\left[\mathbf{b}_1 \mid \mathbf{b}_2 \mid \cdots \mid \mathbf{b}_p\right] = \left[A\mathbf{b}_1 \mid A\mathbf{b}_2 \mid \cdots \mid A\mathbf{b}_p\right]$$

(3) Likewise, we observed in Section 3.1 that if B is partitioned as one row and one column, then

$$AB = \begin{bmatrix} \text{row}_1(A) \\ \text{row}_2(A) \\ \vdots \\ \text{row}_m(A) \end{bmatrix} B = \begin{bmatrix} \text{row}_1(A)B \\ \text{row}_2(A)B \\ \vdots \\ \text{row}_m(A)B \end{bmatrix}$$

(4) Even the scalar row-column rule itself is a special case of the block version:

$$AB = \begin{bmatrix} \text{row}_1(A) \\ \text{row}_2(A) \\ \vdots \\ \text{row}_m(A) \end{bmatrix} [\text{col}_1(B) \quad \text{col}_2(B) \quad \cdots \quad \text{col}_p(B)]$$

$$= \begin{bmatrix} \text{row}_1(A)\text{col}_1(B) & \cdots & \text{row}_1(A)\text{col}_j(B) & \cdots & \text{row}_1(A)\text{col}_p(B) \\ \vdots & & \vdots & & \vdots \\ \text{row}_i(A)\text{col}_1(B) & \cdots & \text{row}_i(A)\text{col}_j(B) & \cdots & \text{row}_i(A)\text{col}_p(B) \\ \vdots & & \vdots & & \vdots \\ \text{row}_m(A)\text{col}_1(B) & \cdots & \text{row}_m(A)\text{col}_j(B) & \cdots & \text{row}_m(A)\text{col}_p(B) \end{bmatrix}$$

In the row-column rule, each *entry* of AB is computed as an *inner product* of the form $\mathbf{u}^T\mathbf{v}$, where the row of a matrix is viewed as the transpose of a column vector. In the next column-row version, AB

itself is expressed as a sum of *outer products* of the form $\mathbf{uv}^T$. Remember that an inner product produces a scalar, while an outer product produces a matrix.

(5) The column-row expansion of AB is the block version of the row-column product in which A has one block row and B has one block column: If A is $m \times n$ and B is $n \times p$, then

$$AB = [\text{col}_1(A) \quad \text{col}_2(A) \quad \cdots \quad \text{col}_n(A)] \begin{bmatrix} \text{row}_1(B) \\ \text{row}_2(B) \\ \vdots \\ \text{row}_n(b) \end{bmatrix}$$

$$= \text{col}_1(A)\cdot\text{row}_1(B) + \cdots + \text{col}_n(A)\cdot\text{row}_n(B)$$

You might say that the row-column rule computes AB as an array of sums while the column-row expansion displays AB as a sum of arrays.

SOLUTIONS TO EXERCISES

1. Apply the row-column rule as if the matrix entries were numbers, but for each product (such as EA below), always write the entry of the left block-matrix on the *left*.

$$\begin{bmatrix} I & 0 \\ E & I \end{bmatrix}\begin{bmatrix} A & B \\ C & D \end{bmatrix} = \begin{bmatrix} IA + 0C & IB + 0D \\ EA + IC & EB + ID \end{bmatrix} = \begin{bmatrix} A & B \\ EA + C & EB + D \end{bmatrix}$$

This must be EA, not AE.

Checkpoint: Notice in Exercises 1 and 3 that $\begin{bmatrix} I & 0 \\ E & I \end{bmatrix}$ and $\begin{bmatrix} 0 & I \\ I & 0 \end{bmatrix}$ act as block-matrix generalizations of elementary matrices. What sort of 2×2 block matrix is the appropriate generalization of an elementary matrix that acts as a scaling operation? (Answer this carefully.)

7. Compute the left side of the equation:

$$\begin{bmatrix} I & 0 & 0 \\ X & I & 0 \\ Y & 0 & I \end{bmatrix}\begin{bmatrix} A_{11} & A_{12} \\ A_{21} & A_{22} \\ A_{31} & A_{32} \end{bmatrix} = \begin{bmatrix} IA_{11} & IA_{12} \\ XA_{11} + IA_{21} & XA_{12} + IA_{22} \\ YA_{11} + IA_{31} & YA_{12} + IA_{32} \end{bmatrix}$$

Set this equal to the right side of the equation:

$$\begin{bmatrix} A_{11} & A_{12} \\ XA_{11} + A_{21} & XA_{12} + A_{22} \\ YA_{11} + A_{31} & YA_{12} + A_{32} \end{bmatrix} = \begin{bmatrix} B_{11} & B_{12} \\ 0 & B_{22} \\ 0 & B_{23} \end{bmatrix}$$

From the (2,1)-entries and (3,1)-entries, since A_{11} is invertible,

$$XA_{11} + A_{21} = 0 \quad \Rightarrow \quad XA_{11} = -A_{21} \quad \Rightarrow \quad X = -A_{21}A_{11}^{-1}$$

$$YA_{11} + A_{31} = 0 \quad \Rightarrow \quad YA_{11} = -A_{31} \quad \Rightarrow \quad Y = -A_{31}A_{11}^{-1}$$

The order of the factors for X and Y is crucial. This type of problem makes a good exam question, because it checks whether you pay attention to the fact that matrix multiplication is generally not commutative.

Finally, since the (2,2)-entries are equal,

$$B_{22} = XA_{12} + A_{22} = -A_{21}A_{11}^{-1}A_{12} + A_{22}$$

13. $A = \begin{bmatrix} A_{11} & A_{12} \\ 0 & A_{22} \end{bmatrix}$. Suppose first that A_{11} and A_{22} are invertible. Example 5 shows what the inverse of A must be *if* it has an inverse, so check whether the candidate for A^{-1} works. Let $C = \begin{bmatrix} A_{11}^{-1} & -A_{11}^{-1}A_{12}A_{22}^{-1} \\ 0 & A_{22}^{-1} \end{bmatrix}$, and compute

$$CA = \begin{bmatrix} A_{11}^{-1} & -A_{11}^{-1}A_{12}A_{22}^{-1} \\ 0 & A_{22}^{-1} \end{bmatrix}\begin{bmatrix} A_{11} & A_{12} \\ 0 & A_{22} \end{bmatrix} = \begin{bmatrix} A_{11}^{-1}A_{11} & A_{11}^{-1}A_{12} - A_{11}^{-1}A_{12}A_{22}^{-1}A_{22} \\ 0 & A_{22}^{-1}A_{22} \end{bmatrix}$$

$$= \begin{bmatrix} I & A_{11}^{-1}A_{12} - A_{11}^{-1}A_{12}I \\ 0 & I \end{bmatrix} = \begin{bmatrix} I & 0 \\ 0 & I \end{bmatrix}$$

Since A is square, this calculation and the IMT imply that A is invertible. (Don't forget this final sentence. Without it, the argument is incomplete.)

Conversely, suppose that A is invertible. Then the calculations in Example 5 show that A_{11} and A_{22} are invertible. (These two sentences are important, too, because the exercise asked you establish an "if and only if" statement. The first paragraph above simply verified the "if" part.)

19. To prove a statement by induction, a good first step is to write the statement that depends on n but exclude the phrase "for all n", and label the statement for reference:

The product of two $n \times n$ *lower triangular matrices is lower triangular.* (*)

Second, verify that the statement is true for $n = 1$. In this particular case, (*) is obviously true, because *every* 1×1 matrix is lower triangular. The "induction step" is next.

Suppose that (*) is true when n is some positive integer k, and consider two $(k+1) \times (k+1)$ matrices B_1 and C_1. Partition B_1 and C_1 as

$$B_1 = \begin{bmatrix} b & \mathbf{0}^T \\ \mathbf{x} & B \end{bmatrix}, \quad C_1 = \begin{bmatrix} c & \mathbf{0}^T \\ \mathbf{y} & C \end{bmatrix}$$

where B and C are $k \times k$ matrices, $\mathbf{x}$ and $\mathbf{y}$ are in $\mathbb{R}^n$, and b and c are scalars. Since B_1 and C_1 are lower triangular, so are B and C. Now,

$$B_1C_1 = \begin{bmatrix} b & \mathbf{0}^T \\ \mathbf{x} & B \end{bmatrix}\begin{bmatrix} c & \mathbf{0}^T \\ \mathbf{y} & C \end{bmatrix} = \begin{bmatrix} bc + \mathbf{0}^T\mathbf{y} & b\mathbf{0}^T + \mathbf{0}^TC \\ \mathbf{x}c + B\mathbf{y} & \mathbf{x}\mathbf{0}^T + BC \end{bmatrix} = \begin{bmatrix} bc & \mathbf{0}^T \\ c\mathbf{x} + B\mathbf{y} & BC \end{bmatrix}$$

Assuming (*) is true for $n = k$, BC must be lower triangular. The form of B_1C_1 shows that it, too, is lower triangular. Thus, the truth of (*) for any positive integer $n = k$ implies the truth of (*) for the next positive integer $k+1$. By the principle of induction, (*) must be true for all $n \geq 1$.

Answer to Checkpoint: The block diagonal matrices $\begin{bmatrix} E & 0 \\ 0 & I \end{bmatrix}$ and $\begin{bmatrix} I & 0 \\ 0 & E \end{bmatrix}$ are obvious choices. Less obvious is the requirement that E be invertible, in order to make these block matrices invertible. (Recall that the invertibility of elementary matrices was essential for the theory in Section 3.2.)

Appendix: The Principle of Induction

Consider a statement "(*)" that depends on a positive integer n, as in Exercise 19. To prove "by induction" that (*) is true for all positive integers, you must prove two things:

(a) Statement (*) is true for $n = 1$.
(b) (The induction step) If (*) is true for any positive integer $n = k$, then (*) is also true for the next integer $n = k + 1$.

A property or axiom of the real number system, called the *principle of mathematical induction*, says that if (a) and (b) are true, then (*) is true for all integers $n \geq 1$. This is reasonable, because if (*) is true for $n = 1$, then (b) shows that (*) is true for $n = 2$. Applying (b) again with $n = 2$, we see that (*) is true for $n = 3$. Applying (b) repeatedly, we see that (*) is true for 2, 3, 4, 5, ...

3.5 MATRIX FACTORIZATIONS

In a sense, Section 3.5 is the most up-to-date section in the text, because matrix factorizations lie at the heart of modern uses of matrix algebra. For instance, they are indispensible for the analysis of computational algorithms and research in parallel processing. The text focuses here on triangular factorizations, but the exercises introduce you to other important factorizations that you may encounter later.

KEY IDEAS

When a matrix A is factored as $A = LU$, the data in A are preprocessed in a way that makes the equation $A\mathbf{x} = \mathbf{b}$ easier to solve. Write $LU\mathbf{x} = \mathbf{b}$, or $L(U\mathbf{x}) = \mathbf{b}$ and let $\mathbf{y} = U\mathbf{x}$. Solve $L\mathbf{y} = \mathbf{b}$ for $\mathbf{y}$ and then solve $U\mathbf{x} = \mathbf{y}$ for $\mathbf{x}$. The two-step process is fast when L and U are triangular.

Finding L and U requires the same number of multiplications and divisions as row reducing A to an echelon form U (about $n^3/3$ operations when A is $n \times n$). After that, L and U are available for solving other equations involving A. The key to finding L is to place entries in L in such a way that the sequence of row operations reducing A to U also reduces L to the identity. In this case, LU must equal A. (See the bottom of page 130.)

The text discusses how to build L when no row interchanges are needed to reduce A to U. In this case, L can be unit lower triangular. An appendix below describes how to build L in permuted unit triangular form when row interchanges are needed (or desired, for numerical reasons).

SOLUTIONS TO EXERCISES

1. $L = \begin{bmatrix} 1 & 0 & 0 \\ -1 & 1 & 0 \\ 2 & -5 & 1 \end{bmatrix}$, $U = \begin{bmatrix} 3 & -7 & -2 \\ 0 & -2 & -1 \\ 0 & 0 & -1 \end{bmatrix}$, $\mathbf{b} = \begin{bmatrix} -7 \\ 5 \\ 2 \end{bmatrix}$. First, solve $L\mathbf{y} = \mathbf{b}$.

$$[L \ \ \mathbf{b}] = \begin{bmatrix} 1 & 0 & 0 & -7 \\ -1 & 1 & 0 & 5 \\ 2 & -5 & 1 & 2 \end{bmatrix} \sim \begin{bmatrix} 1 & 0 & 0 & -7 \\ 0 & 1 & 0 & -2 \\ 0 & -5 & 1 & 16 \end{bmatrix}$$

The only arithmetic was in column 4

$$\sim \begin{bmatrix} 1 & 0 & 0 & -7 \\ 0 & 1 & 0 & -2 \\ 0 & 0 & 1 & 6 \end{bmatrix}, \quad \text{so } \mathbf{y} = \begin{bmatrix} -7 \\ -2 \\ 6 \end{bmatrix}$$

Next, solve $U\mathbf{x} = \mathbf{y}$, using back-substitution (with matrix notation).

$$[U \ \ \mathbf{y}] = \begin{bmatrix} 3 & -7 & -2 & -7 \\ 0 & -2 & -1 & -2 \\ 0 & 0 & -1 & 6 \end{bmatrix} \sim \begin{bmatrix} 3 & -7 & -2 & -7 \\ 0 & -2 & -1 & -2 \\ 0 & 0 & 1 & -6 \end{bmatrix} \sim \begin{bmatrix} 3 & -7 & 0 & -19 \\ 0 & -2 & 0 & -8 \\ 0 & 0 & 1 & -6 \end{bmatrix}$$

$$\sim \begin{bmatrix}3&-7&0&-19\\0&1&0&4\\0&0&1&-6\end{bmatrix} \sim \begin{bmatrix}3&0&0&9\\0&1&0&4\\0&0&1&-6\end{bmatrix} \sim \begin{bmatrix}1&0&0&3\\0&1&0&4\\0&0&1&-6\end{bmatrix}$$

So $\mathbf{x} = (3,4,-6)$.

Checkpoint: Exercise 9 in Section 3.2 shows how to compute $A^{-1}B$ by row reduction. Describe how you could speed up this calculation if you have an LU factorization of A available (and A is invertible).

7. Place the first pivot column of $\begin{bmatrix}2&5\\-3&-4\end{bmatrix}$ into L, after dividing the column by 2 (the pivot), then add 3/2 times row 1 to row 2, yielding U.

$$A = \begin{bmatrix}\textcircled{2}&5\\-3&-4\end{bmatrix} \sim \begin{bmatrix}2&5\\0&\textcircled{7/2}\end{bmatrix} = U, \qquad \begin{bmatrix}\textcircled{2}\\-3\end{bmatrix} \quad \begin{bmatrix}\textcircled{7/2}\end{bmatrix}$$

$$\div 2 \downarrow \qquad \div 7/2 \downarrow$$

$$\begin{bmatrix}1&\\-3/2&1\end{bmatrix}, \quad L = \begin{bmatrix}1&0\\-3/2&1\end{bmatrix}$$

13. $$\begin{bmatrix}\textcircled{1}&3&-5&-3\\-1&-5&8&4\\4&2&-5&-7\\-2&-4&7&5\end{bmatrix} \sim \begin{bmatrix}1&3&-5&-3\\0&\textcircled{-2}&3&1\\0&-10&15&5\\0&2&-3&-1\end{bmatrix} \sim \begin{bmatrix}1&3&-5&-3\\0&-2&3&1\\0&0&0&0\\0&0&0&0\end{bmatrix} = U$$

No more pivots!

$$\begin{bmatrix}\textcircled{1}\\-1\\4\\-2\end{bmatrix} \quad \begin{bmatrix}\textcircled{-2}\\-10\\2\end{bmatrix}$$

Use the last two columns of I_4 to make L unit lower triangular.

$$\div 1 \downarrow \qquad \div -2 \downarrow$$

$$\begin{bmatrix}1&&&\\-1&1&&\\4&5&1&\\-2&-1&0&1\end{bmatrix}, \quad L = \begin{bmatrix}1&0&0&0\\-1&1&0&0\\4&5&1&0\\-2&-1&0&1\end{bmatrix}$$

19. A good answer will require a written paragraph or two. If you have not tried to *write* your answer, do so now, *without reading the solution below*. Explain how you would row reduce $[A \;\; I]$, knowing that A is lower triangular. Your answer to this question should contain some of the ideas shown below, although your wording might be quite different.

Let A be a lower-triangular $n \times n$ matrix with nonzero entries on the diagonal, and consider the augmented matrix $[A \quad I]$.

a. The (1,1)-entry can be scaled to 1 and the entries below it can be changed to 0 by adding multiples of row 1 to the rows below. This affects only the first column of A and the first column of I. So the (2,2)-entry in the new matrix is still nonzero and now is the only nonzero entry of row 2 in the first n columns (because A was lower triangular).

   The (2,2)-entry can be scaled to 1, and the entries below it can be changed to 0 by adding multiples of row 2 to the rows below. This affects only columns 2 and $n+2$ of the augmented matrix. Now the (3,3) entry in A is the only nonzero entry of the third row in the first n columns, so it can be scaled to 1 and then used as a pivot to zero out entries below it. Continuing in this way, A is eventually reduced to I, by scaling each row with a pivot and then using only row operations that add multiples of the pivot row to rows below.

b. The row operations just described only add rows to rows below, so the I on the right in $[A \quad I]$ changes into a lower triangular matrix. By Theorem 7 in Section 3.2, that matrix is A^{-1}.

21. Suppose $A = BC$, with B invertible. Then there exist elementary matrices $E_1, \ldots, E_p$ corresponding to row operations that reduce B to I, in the sense that $E_p \cdots E_1 B = I$. Applying the same sequence of row operations to A amounts to left-multiplying A by the product $E_p \cdots E_1$. By associativity of matrix multiplication,

$$E_p \cdots E_1 A = E_p \cdots E_1 BC = IC = C$$

so the same sequence of row operations reduces A to C.

25. $A = UDV^T$. Since U and V^T are square, the equations $U^TU = I$ and $V^TV = I$ imply that U and V^T are invertible, by the IMT, and hence $U^{-1} = U^T$ and $(V^T)^{-1} = V$.. Since the diagonal entries $\sigma_1, \ldots, \sigma_n$ in D are nonzero, D is invertible, with the inverse of D being the diagonal matrix with $\sigma_1^{-1}, \ldots, \sigma_n^{-1}$ on the diagonal. Thus A is a product of invertible matrices. By Theorem 6, A is invertible and $A^{-1} = (UDV^T)^{-1} = (V^T)^{-1}D^{-1}U^{-1} = VD^{-1}U^T$.

Answer to Checkpoint: If A is an invertible $n \times n$ matrix, with an LU factorization $A = LU$, and if B is $n \times p$, then $A^{-1}B$ can be computed by first row reducing $[L \quad B]$ to a matrix $[I \quad Y]$ for some Y and then reducing $[U \quad Y]$

to $[I \quad A^{-1}B]$. One way to see that this algorithm works is to view $A^{-1}B$ as $[A^{-1}\mathbf{b}_1 \ \cdots \ A^{-1}\mathbf{b}_p]$ and use the LU algorithm to solve simultaneously the set of equations $A\mathbf{x} = \mathbf{b}_1, \ldots, A\mathbf{x} = \mathbf{b}_p$.

Appendix: Permuted LU Factorizations

Any $m \times n$ matrix A admits a factorization $A = LU$, with U in echelon form and L a *permuted unit lower triangular* matrix. That is, L is a matrix such that a permutation (rearrangement) of its rows (using row interchanges) will produce a lower triangular matrix with 1's on the diagonal.

The construction of L and U, illustrated below, depends on first using row replacements to reduce A to a *permuted echelon form* V and then using row interchanges to reduce V to an echelon form U. By watching the reduction of A to V, we can easily construct a permuted unit lower triangular matrix L with the property that the sequence of operations changing A into U also changes L into I. This property will guarantee that $A = LU$. (See the bottom of page 130 in the text.)

The following algorithm reduces any matrix to a permuted echelon form.

1. *Begin with the leftmost nonzero column. Choose any nonzero entry as the pivot. Designate the corresponding row as a pivot row.*
2. *Use row replacements to create zeros above and below the pivot.*
3. *Cover all rows designated as pivot rows, and repeat steps 1 and 2 on the uncovered submatrix, if any, until there are no nonzero entries uncovered.*

This algorithm forces each pivot to be to the right of the preceding pivots; when the rows are rearranged with the pivots in stair-step fashion, all entries below each pivot will be zero. Thus, the algorithm produces a permuted echelon matrix.

For example, choose the (3,1)-entry in the following matrix as the first pivot, and use the pivot to create zeros in the rest of column 1.

$$A = \begin{bmatrix} 1 & -1 & 5 & -8 & -7 \\ -2 & -1 & -4 & 9 & 1 \\ \textcircled{4} & 8 & -4 & 0 & -8 \\ 2 & 3 & 0 & -5 & 3 \end{bmatrix} \sim \begin{bmatrix} 0 & -3 & 6 & -8 & -5 \\ 0 & 3 & -6 & 9 & -3 \\ 4 & 8 & -4 & 0 & -8 \\ 0 & -1 & 2 & -5 & 7 \end{bmatrix} \begin{matrix} \\ \\ \leftarrow \text{1st pivot row} \\ \\ \end{matrix}$$

Row 3 is the 1st pivot row. Choose the (2,2)-entry as the second pivot, and create zeros in the rest of column 2, excluding the first pivot row.

$$= \begin{bmatrix} 0 & -3 & 6 & -8 & -5 \\ 0 & \textcircled{3} & -6 & 9 & -3 \\ 4 & 8 & -4 & 0 & -8 \\ 0 & -1 & 2 & -5 & 7 \end{bmatrix} \sim \begin{bmatrix} 0 & 0 & 0 & 1 & -8 \\ 0 & 3 & -6 & 9 & -3 \\ 4 & 8 & -4 & 0 & -8 \\ 0 & 0 & 0 & -2 & 6 \end{bmatrix} \begin{matrix} \\ \leftarrow \text{2nd pivot row} \\ \leftarrow \text{1st pivot row} \\ \\ \end{matrix}$$

Cover row 2, choose the (4,4)-entry as the pivot. (The row index of the pivot is relative to the original matrix.) Create zeros in the other rows (in the pivot column), excluding the first two pivot rows.

$$= \begin{bmatrix} 0 & 0 & 0 & 1 & -8 \\ 0 & 3 & -6 & 9 & -3 \\ 4 & 8 & -4 & 0 & -8 \\ 0 & 0 & 0 & \textcircled{-2} & 6 \end{bmatrix} \sim \begin{bmatrix} 0 & 0 & 0 & 0 & \textcircled{-5} \\ 0 & \textcircled{3} & -6 & 9 & -3 \\ \textcircled{4} & 8 & -4 & 0 & -8 \\ 0 & 0 & 0 & \textcircled{-2} & 6 \end{bmatrix} \begin{matrix} \leftarrow \text{4th pivot row} \\ \leftarrow \text{2nd pivot row} \\ \leftarrow \text{1st pivot row} \\ \leftarrow \text{3rd pivot row} \end{matrix}$$

Let V denote this permuted echelon form, and permute the rows of V to create an echelon form. The first pivot row goes to the top, the second pivot row goes next, and so on. The resulting echelon matrix U is

$$\begin{bmatrix} 4 & 8 & -4 & 0 & -8 \\ 0 & 3 & -6 & 9 & -3 \\ 0 & 0 & 0 & -2 & 6 \\ 0 & 0 & 0 & 0 & -5 \end{bmatrix} = U$$

The next step is to create L. Watching the reduction of A to V, as each pivot is selected, take the pivot column, and divide the pivot into each entry in the column that is not yet in a pivot row. Place the resulting column into L. At the end, fill the holes in L with zeros.

$$\begin{bmatrix} 1 \\ -2 \\ \textcircled{4} \\ 2 \end{bmatrix} \quad \begin{bmatrix} -3 \\ \textcircled{3} \\ \\ -1 \end{bmatrix} \quad \begin{bmatrix} 1 \\ \\ \\ \textcircled{-2} \end{bmatrix} \quad \begin{bmatrix} \textcircled{-5} \\ \\ \\ \\ \end{bmatrix}$$

$$\begin{matrix} \div 4 & \div 3 & \div -2 & \div -5 \\ \downarrow & \downarrow & \downarrow & \downarrow \end{matrix}$$

$$\begin{bmatrix} 1/4 & -1 & -1/2 & 1 \\ -1/2 & 1 & & \\ 1 & & & \\ 1/2 & -1/3 & 1 & \end{bmatrix}, \qquad L = \begin{bmatrix} 1/4 & -1 & -1/2 & 1 \\ -1/2 & 1 & 0 & 0 \\ 1 & 0 & 0 & 0 \\ 1/2 & -1/3 & 1 & 0 \end{bmatrix}$$

At this point, L is constructed so the operations that reduce A to V also reduce L to a permuted identity matrix. Since the pivots in L are in exactly the same *rows* as in A, the sequence of row interchanges that reduces V to U also reduces the permuted identity matrix to I. Thus, the full sequence of operations that reduces A to U also reduces L to I, so that $A = LU$.

The next example illustrates what to do when V has one or more rows of zeros. The matrix is from the Practice Problem for Section 3.5. For the reduction of A to V, pivots were chosen to have the largest possible magnitude (the choice used for "partial pivoting"). Of course, other pivots could have been selected.

$$A = \begin{bmatrix} 2 & -4 & -2 & 3 \\ \textcircled{6} & -9 & -5 & 8 \\ 2 & -7 & -3 & 9 \\ 4 & -2 & -2 & -1 \\ -6 & 3 & 3 & 4 \end{bmatrix} \sim \begin{bmatrix} 0 & -1 & -1/3 & 1/3 \\ 6 & -9 & -5 & 8 \\ 0 & -4 & -4/3 & 19/3 \\ 0 & 4 & 4/3 & -19/3 \\ 0 & \textcircled{-6} & -2 & 12 \end{bmatrix} \sim \begin{bmatrix} 0 & 0 & 0 & -5/3 \\ 6 & -9 & -5 & 8 \\ 0 & 0 & 0 & -5/3 \\ 0 & 0 & 0 & \textcircled{5/3} \\ 0 & -6 & -2 & 12 \end{bmatrix}$$

$$\sim V = \begin{bmatrix} 0 & 0 & 0 & 0 \\ \textcircled{6} & -9 & -5 & 8 \\ 0 & 0 & 0 & 0 \\ 0 & 0 & 0 & \textcircled{5/3} \\ 0 & \textcircled{-6} & -2 & 12 \end{bmatrix} \begin{matrix} \\ \leftarrow \text{1st pivot row} \\ \\ \leftarrow \text{3rd pivot row} \\ \leftarrow \text{2nd pivot row} \end{matrix} \qquad \sim U = \begin{bmatrix} 6 & -9 & -5 & 8 \\ 0 & -6 & -2 & 12 \\ 0 & 0 & 0 & 5/3 \\ 0 & 0 & 0 & 0 \\ 0 & 0 & 0 & 0 \end{bmatrix}$$

The first three columns of L come from the three pivot columns above.

$$\begin{bmatrix} 2 \\ \textcircled{6} \\ 2 \\ 4 \\ -6 \end{bmatrix} \quad \begin{bmatrix} -1 \\ \\ -4 \\ 4 \\ \textcircled{-6} \end{bmatrix} \quad \begin{bmatrix} -5/3 \\ \\ -5/3 \\ \textcircled{5/3} \\ \\ \end{bmatrix}$$

$$\div 6 \qquad \div -6 \qquad \div 5/3$$
$$\downarrow \qquad \downarrow \qquad \downarrow$$

$$\begin{bmatrix} 1/3 & 1/6 & -1 & & \\ 1 & & & & \\ 1/3 & 2/3 & -1 & & \\ 2/3 & -2/3 & 1 & & \\ -1 & 1 & & & \end{bmatrix} \begin{matrix} \\ \leftarrow \text{1st pivot row} \\ \\ \leftarrow \text{3rd pivot row} \\ \leftarrow \text{2nd pivot row} \end{matrix}$$

The matrix L needs two more columns. Use columns 1 and 3 of the 5×5 identity matrix to place 1's in the "nonpivot" rows 1 and 3. Fill in the remaining holes with zeros.

$$\begin{bmatrix} 1/3 & 1/6 & -1 & 1 & 0 \\ \textcircled{1} & 0 & 0 & 0 & 0 \\ 1/3 & 2/3 & -1 & 0 & 1 \\ 2/3 & -2/3 & \textcircled{1} & 0 & 0 \\ -1 & \textcircled{1} & 0 & 0 & 0 \end{bmatrix} = L \sim \begin{bmatrix} 1 & 0 & 0 & 0 & 0 \\ -1 & 1 & 0 & 0 & 0 \\ 2/3 & -2/3 & 1 & 0 & 0 \\ 1/3 & 1/6 & -1 & 1 & 0 \\ 1/3 & 2/3 & -1 & 0 & 1 \end{bmatrix}$$

Row reduction of L using only row replacements will produce a permuted identity matrix, with 1's in the "pivot rows" 2, 5, and 4. Moving these rows into rows 1,2, and 3 requires the same operations as reducing V to U. If a further row interchange on the permuted identity is required, it will involve the bottom two rows, which came from the "nonpivot" rows 1 and 3. A corresponding interchange of the bottom two rows of U has no effect on U (and the product LU is unaffected). As a result, L is reduced to I by the same operations that reduce A to V and then to U. Check that $A = LU$.

MATLAB Row reduction of *A* using the command **gauss(A,p)**, described in the MATLAB notes for Section 3.2, will produce the intermediate matrices needed for an LU factorization of *A*. You can try this on the matrix in Example 2—stored in Exercise 31 on your data disk. The matrices in (4) on page 131 in the text are produced by the commands

```
U = gauss(A,1)      A matrix with 0's below the first pivot
U = gauss(U,2)      A matrix with 0's below pivots 1 and 2
U = gauss(U,3)      The echelon form, U
```

You can copy the information from the screen onto your paper, and divide by the pivot entries to produce *L* as in the text. (For text exercises, the pivots are integers and so are displayed accurately.)

To actually construct *L* in MATLAB, you can extract parts of columns of the matrix as it is being reduced. As an example, the commands that follow apply to the 4×5 matrix *A* in Example 2. First, create an *L* of the correct size, **L = eye(4)** , and enter the command:

```
L(:,1) = A(:,1)/A(1,1)          Fill the first column of L.
```

Then, after cleaning out the first column with **U = gauss(A,1)**, you can fill in rows 2 to 4 of column 2 in *L* with rows 2 to 4 of column 2 in *U*, divided by the (2,2)-pivot:

```
L(2:4,2) = U(2:4,2)/U(2,2)      Recall that 2:4 means "2 to 4".
```

After the next gauss operation, **U = gauss(U,2)**, the third pivot of the "new" *U* is in the (3,4)-position. So the *third* column of *L* (from the diagonal down) comes from the *fourth* column of *U*, using rows 3 and 4:

```
L(3:4,3) = U(3:4,4)/U(3,4)
```

and so on. In general, if the *r*th pivot appears in the (*r*,*s*)-position, and if the partially reduced *U* has *m* rows, then the rth column of *L*, from the diagonal down, is filled by the command:

```
L(r:m,r) = U(r:m,s)/U(r,s)
```

To construct a permuted LU factorization, use **U = gauss(U,r,v)** , where *r* is the row index of the pivot and **v** is a row vector that lists the rows to be changed by replacement operations. For example, if *A* has 5 rows and the first pivot is in row 4, use **U = gauss(A,4,[1 2 3 5])** . If the next pivot is in row 2, use **U = gauss(U,2,[1 3 5])** . To build the permuted matrix *L*, use full columns from *A* or the partially reduced *U*, divided by the pivots. Then change entries to zero if they are in a row already selected as a "pivot row."

The MATLAB command **[L U] = lu(A)** produces a permuted LU factorization for any square matrix *A*, but it does not handle the general case.

3.6 ITERATIVE SOLUTIONS OF LINEAR SYSTEMS

Some knowledge of iterative methods for solving linear systems is important for anyone who uses linear algebra. If your course does not have time to cover this material, you might at least glance at the discussion. Then, if you need to solve a large sparse system later in your career, you may remember (perhaps vaguely) the ideas here, and you can review this section before setting out to use appropriate computer software.

KEY IDEAS

Both iterative algorithms described here are based on writing $A\mathbf{x} = \mathbf{b}$ in the form $(M - N)\mathbf{x} = \mathbf{b}$, where M is readily invertible, and then writing $M\mathbf{x} = N\mathbf{x} + \mathbf{b}$. Given an initial estimate $\mathbf{x}^{(0)}$ for the solution, the recursion

$$M\mathbf{x}^{(k+1)} = N\mathbf{x}^{(k)} + \mathbf{b} \quad (k = 0, 1, \dots) \tag{*}$$

defines a sequence of vectors that may converge to the desired solution.

For Jacobi's method, M is the diagonal matrix formed from the diagonal entries in A; for the Gauss-Seidel method, M is the lower triangular part of A. In both cases, of course, $N = M - A$.

SOLUTIONS TO EXERCISES

1. $A = \begin{bmatrix} 4 & 1 \\ -1 & 5 \end{bmatrix}$, $\mathbf{b} = \begin{bmatrix} 7 \\ -7 \end{bmatrix}$, $M = \begin{bmatrix} 4 & 0 \\ 0 & 5 \end{bmatrix}$, $N = \begin{bmatrix} 4 & 0 \\ 0 & 5 \end{bmatrix} - \begin{bmatrix} 4 & 1 \\ -1 & 5 \end{bmatrix} = \begin{bmatrix} 0 & -1 \\ 1 & 0 \end{bmatrix}$.

If working by hand, the recursion (*) above can be simplified to

$$\mathbf{x}^{(k+1)} = M^{-1}(N\mathbf{x}^{(k)} + \mathbf{b}) = \begin{bmatrix} .25 & 0 \\ 0 & .20 \end{bmatrix}\left(\begin{bmatrix} 0 & -1 \\ 1 & 0 \end{bmatrix}\mathbf{x}^{(k)} + \begin{bmatrix} 7 \\ -7 \end{bmatrix}\right)$$

$$= \begin{bmatrix} 0 & -.25 \\ .2 & 0 \end{bmatrix}\mathbf{x}^{(k)} + \begin{bmatrix} 1.75 \\ -1.40 \end{bmatrix}$$

Then, with $\mathbf{x}^{(0)} = \mathbf{0}$,

$$\mathbf{x}^{(1)} = \begin{bmatrix} 0 & -.25 \\ .2 & 0 \end{bmatrix}\mathbf{0} + \begin{bmatrix} 1.75 \\ -1.40 \end{bmatrix} = \begin{bmatrix} 1.75 \\ -1.40 \end{bmatrix}$$

$$\mathbf{x}^{(2)} = \begin{bmatrix} 0 & -.25 \\ .2 & 0 \end{bmatrix}\begin{bmatrix} 1.75 \\ -1.40 \end{bmatrix} + \begin{bmatrix} 1.75 \\ -1.40 \end{bmatrix} = \begin{bmatrix} .35 \\ .35 \end{bmatrix} + \begin{bmatrix} 1.75 \\ -1.40 \end{bmatrix} = \begin{bmatrix} 2.10 \\ -1.05 \end{bmatrix}$$

$$\mathbf{x}^{(3)} = \begin{bmatrix} 0 & -.25 \\ .2 & 0 \end{bmatrix}\begin{bmatrix} 2.10 \\ -1.05 \end{bmatrix} + \begin{bmatrix} 1.75 \\ -1.40 \end{bmatrix}$$

Using MATLAB,

$$\mathbf{x}^{(4)} = \begin{bmatrix}1.9950\\-.9975\end{bmatrix}, \quad \mathbf{x}^{(5)} = \begin{bmatrix}1.9994\\-1.0010\end{bmatrix}, \quad \mathbf{x}^{(6)} = \begin{bmatrix}2.0003\\-1.0001\end{bmatrix}$$

To four decimal places,

$$\mathbf{x}^{(5)} - \mathbf{x}^{(4)} = \begin{bmatrix}.0044\\-.0035\end{bmatrix}, \quad \mathbf{x}^{(6)} - \mathbf{x}^{(5)} = \begin{bmatrix}.0009\\.0009\end{bmatrix}$$

Thus $\mathbf{x}^{(6)}$ is the first vector in the sequence $\{\mathbf{x}^{(k)}\}$ whose entries are within .001 of the entries in its predecessor.

7. $A = \begin{bmatrix}3 & 1 & 0\\-1 & -5 & 2\\0 & 3 & 7\end{bmatrix}$, $\mathbf{b} = \begin{bmatrix}11\\15\\17\end{bmatrix}$. For Gauss-Seidel, $M = \begin{bmatrix}3 & 0 & 0\\-1 & -5 & 0\\0 & 3 & 7\end{bmatrix}$, and

$$N = \begin{bmatrix}3 & 0 & 0\\-1 & -5 & 0\\0 & 3 & 7\end{bmatrix} - \begin{bmatrix}3 & 1 & 0\\-1 & -5 & 2\\0 & 3 & 7\end{bmatrix} = \begin{bmatrix}0 & -1 & 0\\0 & 0 & -2\\0 & 0 & 0\end{bmatrix}$$

If working by hand, write the recursion (*) as $\mathbf{x}^{(k+1)} = M^{-1}N\mathbf{x}^{(k)} + M^{-1}\mathbf{b}$. This will avoid solving a 3×3 system for each iteration. The fastest way to compute $M^{-1}N$ and $M^{-1}\mathbf{b}$ is to reduce the augmented matrix $[M \;\; N \;\; \mathbf{b}]$ to a matrix of the form $[I \;\; M^{-1}N \;\; M^{-1}\mathbf{b}]$. See Exercise 9 in Section 2, with the matrix B replaced by $[N \;\; \mathbf{b}]$.

$$\left[\begin{array}{ccc|ccc|c}3 & 0 & 0 & 0 & -1 & 0 & 11\\-1 & -5 & 0 & 0 & 0 & -2 & 15\\0 & 3 & 7 & 0 & 0 & 0 & 17\end{array}\right] \sim \begin{bmatrix}1 & 0 & 0 & 0 & -1/3 & 0 & 11/3\\-1 & -5 & 0 & 0 & 0 & -2 & 15\\0 & 3 & 7 & 0 & 0 & 0 & 17\end{bmatrix}$$

$$\sim \begin{bmatrix}1 & 0 & 0 & 0 & -1/3 & 0 & 11/3\\0 & -5 & 0 & 0 & -1/3 & -2 & 56/3\\0 & 3 & 7 & 0 & 0 & 0 & 17\end{bmatrix} \sim \cdots$$

$$\sim \begin{bmatrix}1 & 0 & 0 & 0 & -1/3 & 0 & 11/3\\0 & 1 & 0 & 0 & 1/15 & 2/5 & -56/15\\0 & 0 & 1 & 0 & -1/35 & -6/35 & 141/35\end{bmatrix}$$

$$M^{-1}N = \begin{bmatrix}0 & -1/3 & 0\\0 & 1/15 & 2/5\\0 & -1/35 & -6/35\end{bmatrix} = \frac{1}{105}\begin{bmatrix}0 & -35 & 0\\0 & 7 & 42\\0 & -3 & -18\end{bmatrix} \approx \begin{bmatrix}0 & -.33333 & 0\\0 & .06667 & .40000\\0 & -.02857 & -.17143\end{bmatrix}$$

$$M^{-1}\mathbf{b} = \begin{bmatrix}11/3\\-56/15\\141/35\end{bmatrix} = \frac{1}{105}\begin{bmatrix}385\\-392\\423\end{bmatrix} \approx \begin{bmatrix}3.66667\\-3.73333\\4.02857\end{bmatrix}$$

The first two iterations, by hand (with $\mathbf{x}^{(0)} = \mathbf{0}$), are

$$\mathbf{x}^{(1)} = M^{-1}N\mathbf{0} + M^{-1}\mathbf{b} = \begin{bmatrix}0\\0\\0\end{bmatrix} + \frac{1}{105}\begin{bmatrix}385\\-392\\423\end{bmatrix} = \frac{1}{105}\begin{bmatrix}385\\-392\\423\end{bmatrix}$$

$$\mathbf{x}^{(2)} = M^{-1}N(M^{-1}\mathbf{b}) + M^{-1}\mathbf{b} = (M^{-1}N + I)M^{-1}\mathbf{b}$$ Using matrix algebra to simplify the calculations

$$= \frac{1}{105}\begin{bmatrix}105 & -35 & 0\\0 & 112 & 42\\0 & -3 & 87\end{bmatrix}\left(\frac{1}{105}\right)\begin{bmatrix}385\\-392\\423\end{bmatrix} = \frac{1}{105^2}\begin{bmatrix}54145\\-26138\\37977\end{bmatrix} \approx \begin{bmatrix}4.9111\\-2.3708\\3.4446\end{bmatrix}$$

Your answers may vary slightly if you used decimal approximations throughout your work. Next, using MATLAB (thankfully!),

k	2	3	4	5	6
$\mathbf{x}^{(k)}$	$\begin{bmatrix}4.9111\\-2.3708\\3.4446\end{bmatrix}$	$\begin{bmatrix}4.4569\\-2.5135\\3.5058\end{bmatrix}$	$\begin{bmatrix}4.5045\\-2.4986\\3.4994\end{bmatrix}$	$\begin{bmatrix}4.49953\\-2.50015\\3.50006\end{bmatrix}$	$\begin{bmatrix}4.50005\\-2.49998\\3.49999\end{bmatrix}$

To four decimal places,

$$\mathbf{x}^{(5)} - \mathbf{x}^{(4)} = \begin{bmatrix}-.0050\\-.0016\\.0007\end{bmatrix}, \quad \mathbf{x}^{(6)} - \mathbf{x}^{(5)} = \begin{bmatrix}.0005\\.0002\\-.0001\end{bmatrix}$$

Thus $\mathbf{x}^{(6)}$ is the first vector in the sequence $\{\mathbf{x}^{(k)}\}$ whose entries are within .001 of the entries in its predecessor.

13. $A = \begin{bmatrix}4 & -2\\-5 & 4\end{bmatrix}$, $\mathbf{b} = \begin{bmatrix}10\\-2\end{bmatrix}$, $M = \begin{bmatrix}4 & 0\\-5 & 4\end{bmatrix}$, $N = \begin{bmatrix}0 & 2\\0 & 0\end{bmatrix}$

Some computational aid such as MATLAB is essential here. To help you check your work, if necessary, the results of the first few computations are shown below. (See the text for the final answer.)

k	0	1	2	3	4	5	6	7	8
$\mathbf{x}^{(k)}$	$\begin{bmatrix}0\\0\end{bmatrix}$	$\begin{bmatrix}2.500\\2.625\end{bmatrix}$	$\begin{bmatrix}3.8\\4.3\end{bmatrix}$	$\begin{bmatrix}4.6\\5.3\end{bmatrix}$	$\begin{bmatrix}5.1\\5.9\end{bmatrix}$	$\begin{bmatrix}5.5\\6.3\end{bmatrix}$	$\begin{bmatrix}5.7\\6.6\end{bmatrix}$	$\begin{bmatrix}5.8\\6.7\end{bmatrix}$	$\begin{bmatrix}5.9\\6.8\end{bmatrix}$

17. c. Suppose that $\mathbf{x}^{(0)} = \mathbf{0}$ and $\mathbf{x}^{(k)}$ satisfies $\mathbf{x}^{(k+1)} = C\mathbf{x}^{(k)} + \mathbf{d}$, for $k = 1,2,\ldots.$ To prove by induction that

$$\mathbf{x}^{(k)} = \mathbf{d} + C\mathbf{d} + \cdots + C^{k-1}\mathbf{d} \qquad (k = 1,2,\ldots) \qquad (*)$$

observe that (*) holds for $k = 1$, because by the recursion relation, $\mathbf{x}^{(1)} = C\mathbf{x}^{(0)} + \mathbf{d} = C\mathbf{0} + \mathbf{d} = \mathbf{d}$. Also, (*) holds for $k = 2$, because $\mathbf{x}^{(2)} = C\mathbf{x}^{(1)} + \mathbf{d} = \mathbf{d} + C\mathbf{d}$. Now suppose that (*) holds for some specific $k \geq 1$. Substitute the formula (*) for $\mathbf{x}^{(k)}$ into the recursion relation:

$$\begin{aligned}\mathbf{x}^{(k+1)} &= C\mathbf{x}^{(k)} + \mathbf{d} = C(\mathbf{d} + C\mathbf{d} + \cdots + C^{k-1}\mathbf{d}) + \mathbf{d} \\ &= C\mathbf{d} + C^2\mathbf{d} + \cdots + C^k\mathbf{d} + \mathbf{d}\end{aligned}$$

Thus the truth of (*) for k implies the truth of (*) for $k + 1$. By the principle of induction, (*) is true for all $k \geq 1$.

MATLAB The command **M = diag(diag(A))** creates a diagonal matrix from the diagonal entries of a matrix A. (The command **diag(A)** produces a column vector that lists the diagonal entries of A.) To create the lower triangular part of A for the Gauss-Seidel Method, use the command **M = tril(A)**.

The Jacobi and Gauss-Seidel methods use the same MATLAB commands, except for the construction of M. Enter **N = M - A** and make sure the initial data is in the vector **x**. To produce the "next" **x**, you must solve the equation $M\mathbf{x}^{(1)} = N\mathbf{x} + \mathbf{b}$ for $\mathbf{x}^{(1)}$. The correct way to do this is to use MATLAB's "backslash" operation: $\mathbf{x}^{(1)} = M\backslash(N*\mathbf{x} + \mathbf{b})$. (This operation automates the row reduction that you have done in steps until now.) However, instead of assigning the solution to $\mathbf{x}^{(1)}$, you should put the solution back into **x**, so you can repeat the operation *with exactly the same symbols*. That is, the command

```
x = M\(N*x + b); x'        To save space, x is displayed horizontally
```

is used over and over, creating the sequence $\mathbf{x}^{(k)}$. (The semicolon suppresses the display of the column vector **x**.) These vectors can be recorded on paper as they appear. You only type the command once. After that, use the up-arrow to recall the command, and press <Enter>.

If you wish to monitor how close the approximations are to each other, use the following sequence of commands (repeatedly):

```
t = M\(N*x + b); t'        t is a temporary storage for the new x
change = t' - x'           Compare t with the old x
x = t;                     Update x with new values; don't display them
```

3.7 THE LEONTIEF INPUT-OUTPUT MODEL

If you are in economics, you definitely will need the material in this section for later work. Although most of the discussion concerns economics, the formula for the inverse of $I - C$ is used in a variety of applications.

KEY IDEAS

The power of Leontief's model of the economy is that it compresses hundreds of equations in hundreds of variables into the simple matrix equation $(I - C)\mathbf{x} = \mathbf{d}$. You should know how to construct the consumption matrix C and know the algebra that leads from the matrix equation $\mathbf{x} = C\mathbf{x} + \mathbf{d}$ to its solution $\mathbf{x} = (I - C)^{-1}\mathbf{d}$, under the assumption that the column sums of C are less than one. You may need to know the formula (8) for $(I - C)^{-1}$ on page 146. (Check with your instructor.)

SOLUTIONS TO EXERCISES

1. Fill in C one column at a time, since each column is a unit consumption vector for one sector. Make sure that the order of the sectors is the same for the rows and columns of C. From the way the data are presented, we use the order: manufacturing, agriculture, and services. Read the sentences carefully, to get the data arranged correctly.

Purchased from:	Unit consumption vectors Manuf.	Agric.	Serv.
Manufacturing	.10	.60	.60
Agriculture	.30	.20	.00
Services	.30	.10	.10

   The intermediate demands created by a production vector $\mathbf{x}$ are given by $C\mathbf{x}$. If agriculture plans to produce 100 units (and the other sectors plan to produce nothing), then the intermediate demand is

$$C\mathbf{x} = \begin{bmatrix} .10 & .60 & .60 \\ .30 & .20 & .00 \\ .30 & .10 & .10 \end{bmatrix}\begin{bmatrix} 0 \\ 100 \\ 0 \end{bmatrix} = \begin{bmatrix} 60 \\ 20 \\ 10 \end{bmatrix}$$

7. $C = \begin{bmatrix} .0 & .5 \\ .6 & .2 \end{bmatrix}$, $\mathbf{d} = \begin{bmatrix} 50 \\ 30 \end{bmatrix}$. Let $\mathbf{d}_1 = \begin{bmatrix} 1 \\ 0 \end{bmatrix}$, the demand for 1 unit of output of sector 1.

   a. The production required to satisfy the demand $\mathbf{d}_1$ is the vector $\mathbf{x}_1$ such that $(I - C)\mathbf{x}_1 = \mathbf{d}_1$, namely, $\mathbf{x}_1 = (I - C)^{-1}\mathbf{d}_1$. From Exercise 5,

$$I - C = \begin{bmatrix} 1 & -.5 \\ -.6 & .8 \end{bmatrix} \quad \text{and} \quad (I - C)^{-1} = \begin{bmatrix} 1.6 & 1 \\ 1.2 & 2 \end{bmatrix}$$

so

$$\mathbf{x}_1 = \begin{bmatrix} 1.6 & 1 \\ 1.2 & 2 \end{bmatrix}\begin{bmatrix} 1 \\ 0 \end{bmatrix} = \begin{bmatrix} 1.6 \\ 1.2 \end{bmatrix}$$

b. For the final demand $\mathbf{d}_2 = \begin{bmatrix} 51 \\ 30 \end{bmatrix}$, the corresponding production $\mathbf{x}_2$ is given by

$$\mathbf{x}_2 = (I - C)^{-1}\mathbf{d}_2 = \begin{bmatrix} 1.6 & 1 \\ 1.2 & 2 \end{bmatrix}\begin{bmatrix} 51 \\ 30 \end{bmatrix} = \begin{bmatrix} 111.6 \\ 121.2 \end{bmatrix}$$

c. From Exercise 5, the production $\mathbf{x}$ corresponding to the demand $\mathbf{d}$ is given by $\mathbf{x} = \begin{bmatrix} 110 \\ 120 \end{bmatrix}$. Observe from (a) and (b) that $\mathbf{x}_2 = \mathbf{x} + \mathbf{x}_1$. Also, as pointed out in the text, $\mathbf{d}_2 = \mathbf{d} + \mathbf{d}_1$. The sum of the production vectors $\mathbf{x}$ and $\mathbf{x}_1$ gives the production needed to satisfy the sum of the demands $\mathbf{d}$ and $\mathbf{d}_1$. This is expressing the *linearity* between final demand and production. This relation is true in general, because

$$\begin{aligned} \mathbf{x}_2 &= (I - C)^{-1}\mathbf{d}_2 = (I - C)^{-1}(\mathbf{d} + \mathbf{d}_1) \\ &= (I - C)^{-1}\mathbf{d} + (I - C)^{-1}\mathbf{d}_1 \\ &= \mathbf{x} + \mathbf{x}_1 \end{aligned}$$

11. Begin with the price equation, $\mathbf{p} = C^T\mathbf{p} + \mathbf{v}$. Taking transposes yields

$$\mathbf{p}^T = (C^T\mathbf{p})^T + \mathbf{v}^T = \mathbf{p}^T C^{TT} + \mathbf{v}^T = \mathbf{p}^T C + \mathbf{v}^T$$

and right multiplication by $\mathbf{x}$ produces

$$\mathbf{p}^T\mathbf{x} = \mathbf{p}^T C\mathbf{x} + \mathbf{v}^T\mathbf{x} \qquad (1)$$

Using the production equation $\mathbf{x} = C\mathbf{x} + \mathbf{d}$, compute

$$\mathbf{p}^T\mathbf{x} = \mathbf{p}^T(C\mathbf{x} + \mathbf{d}) = \mathbf{p}^T C\mathbf{x} + \mathbf{p}^T\mathbf{d} \qquad (2)$$

Equating the two expressions (1) and (2) for $\mathbf{p}^T\mathbf{x}$ yields

$$\mathbf{p}^T C\mathbf{x} + \mathbf{p}^T\mathbf{d} = \mathbf{p}^T C\mathbf{x} + \mathbf{v}^T\mathbf{x}$$

which implies that $\mathbf{p}^T\mathbf{d} = \mathbf{v}^T\mathbf{x}$, as desired.

3.8 APPLICATIONS TO COMPUTER GRAPHICS

According to my students over the past few years, this section is one of the most interesting application sections in the text, because it shows how matrix calculations, performed millions of times per second, can create the illusion of 3D-motion on a computer screen or in a movie theater. Of course, one short section cannot begin to indicate the vast scope of computer graphics. I encourage you to look at the book by Foley, et al., referenced in your text. Chapters 5, 6, and 11 are filled with matrices! The rest of the 1100 pages in the book contains lots of interesting mathematics, detailed discussions of computer algorithms, and scores of spectacular (in some cases, almost unbelievable) color plates.

KEY IDEAS

A graphical object, represented by a number of key points, can be stored in a data matrix D, one column for each key point. When a linear transformation, determined by a matrix A, acts on the object, the transformed object is determined by the data matrix AD, because A acts separately on each column of D.

Homogeneous coordinates are needed to make translation act as a linear transformation and to provide perspective projections that give the illusion of three dimensions. The matrix for these transformations has the form

$$\begin{bmatrix} A & \mathbf{p} \\ \mathbf{q}^T & r \end{bmatrix}$$

For 2D-graphics, A is a 2×2 matrix, $\mathbf{q}^T = [0 \quad 0]$, and $\mathbf{p}$ is a translation vector. For 3D-graphics, A is 3×3, $\mathbf{q}$ is associated with a perspective transformation (if any), and $\mathbf{p}$ is a translation vector. For simplicity, the only perspective transformations considered have their center of projection at $(0,0,d)$, and $\mathbf{q}^T = [0 \quad 0 \quad -1/d]$. In fact, $d = 10$ in Example 8 and Exercises 19 and 20.

SOLUTIONS TO EXERCISES

1. From Example 5, the matrix $\begin{bmatrix} 1 & .25 & 0 \\ 0 & 1 & 0 \\ 0 & 0 & 1 \end{bmatrix}$ has the same effect on homogeneous coordinates for $\mathbb{R}^2$ that the matrix $\begin{bmatrix} 1 & .25 \\ 0 & 1 \end{bmatrix}$ of Example 2 has on ordinary vectors in $\mathbb{R}^2$. Partitioned matrix notation explains why this is true. Let A be a 2×2 matrix. The following diagram shows that the action of $\begin{bmatrix} A & \mathbf{0} \\ \mathbf{0}^T & 1 \end{bmatrix}$ on $\begin{bmatrix} \mathbf{x} \\ 1 \end{bmatrix}$ corresponds to the action of A on $\mathbf{x}$.

$$\begin{array}{ccc} \mathbf{x} & \longmapsto & A\mathbf{x} \\ \updownarrow & & \updownarrow \\ \begin{bmatrix} \mathbf{x} \\ 1 \end{bmatrix} & \longmapsto \begin{bmatrix} A & \mathbf{0} \\ \mathbf{0}^T & 1 \end{bmatrix}\begin{bmatrix} \mathbf{x} \\ 1 \end{bmatrix} = \begin{bmatrix} A\mathbf{x} + \mathbf{0}\cdot 1 \\ \mathbf{0}^T\mathbf{x} + 1\cdot 1 \end{bmatrix} = & \begin{bmatrix} A\mathbf{x} \\ 1 \end{bmatrix} \end{array}$$

Coordinates in $\mathbb{R}^2$

Homogeneous coordinates

7. A 60% rotation about the origin in $\mathbb{R}^2$ is given by $\begin{bmatrix} \cos 60^\circ & -\sin 60^\circ \\ \sin 60^\circ & \cos 60^\circ \end{bmatrix} =$ $\begin{bmatrix} 1/2 & -\sqrt{3}/2 \\ \sqrt{3}/2 & 1/2 \end{bmatrix}$, so the 3×3 matrix for rotation about $\begin{bmatrix} 6 \\ 8 \end{bmatrix}$ is

$$\begin{bmatrix} 1 & 0 & 6 \\ 0 & 1 & 8 \\ 0 & 0 & 1 \end{bmatrix}\begin{bmatrix} 1/2 & -\sqrt{3}/2 & 0 \\ \sqrt{3}/2 & 1/2 & 0 \\ 0 & 0 & 1 \end{bmatrix}\begin{bmatrix} 1 & 0 & -6 \\ 0 & 1 & -8 \\ 0 & 0 & 1 \end{bmatrix}$$

Finally, translate back — Then, rotate about the origin — First, translate by $-\mathbf{p}$

$$= \begin{bmatrix} 1 & 0 & 6 \\ 0 & 1 & 8 \\ 0 & 0 & 1 \end{bmatrix}\begin{bmatrix} 1/2 & -\sqrt{3}/2 & -3+4\sqrt{3} \\ \sqrt{3}/2 & 1/2 & -4-3\sqrt{3} \\ 0 & 0 & 1 \end{bmatrix} = \begin{bmatrix} 1/2 & -\sqrt{3}/2 & 3+4\sqrt{3} \\ \sqrt{3}/2 & 1/2 & 4-3\sqrt{3} \\ 0 & 0 & 1 \end{bmatrix}$$

13. The answer is given in the text. Notice that the order of the transformations is important. If the translation is done first (that is, if the matrix for the translation is on the right), then

$$\begin{bmatrix} A & \mathbf{0} \\ \mathbf{0}^T & 1 \end{bmatrix}\begin{bmatrix} I & \mathbf{p} \\ \mathbf{0}^T & 1 \end{bmatrix} = \begin{bmatrix} AI + \mathbf{0}\mathbf{0}^T & A\mathbf{p} + \mathbf{0}\cdot 1 \\ \mathbf{0}^T I + 1\mathbf{0}^T & \mathbf{0}^T\mathbf{p} + 1\cdot 1 \end{bmatrix} = \begin{bmatrix} A & A\mathbf{p} \\ \mathbf{0}^T & 1 \end{bmatrix} \neq \begin{bmatrix} A & \mathbf{p} \\ \mathbf{0}^T & 1 \end{bmatrix}$$

Here, $\mathbf{0}^T$ is a zero row vector, and so the outer product $\mathbf{0}\mathbf{0}^T$ is a zero matrix. Note: The statement of Exercise 13 and its answer in the first and second printings of the text used only "0" instead of $\mathbf{0}$ and $\mathbf{0}^T$, which should cause no confusion but was incorrect.

19. The matrix P for the perspective transformation with center of projection at (0,0,10) and the data matrix D using homogeneous coordinates are shown below. The data matrix for the image of the triangle is PD:

$$PD = \begin{bmatrix} 1 & 0 & 0 & 0 \\ 0 & 1 & 0 & 0 \\ 0 & 0 & 0 & 0 \\ 0 & 0 & -.1 & 1 \end{bmatrix} \begin{bmatrix} 4.2 & 6 & 2 \\ 1.2 & 4 & 2 \\ 4 & 2 & 6 \\ 1 & 1 & 1 \end{bmatrix} = \begin{bmatrix} 4.2 & 6 & 2 \\ 1.2 & 4 & 2 \\ 0 & 0 & 0 \\ .6 & .8 & .4 \end{bmatrix}$$

The $\mathbb{R}^3$ coordinates of the image points come from the top three entries in each column, divided by the corresponding entries in the fourth row.

$$\begin{bmatrix} 4.2/.6 & 6/.8 & 2/.4 \\ 1.2/.6 & 4/.8 & 2/.4 \\ 0 & 0 & 0 \end{bmatrix} = \begin{bmatrix} 7 & 7.5 & 5 \\ 2 & 5 & 5 \\ 0 & 0 & 0 \end{bmatrix}$$

CHAPTER 3 GLOSSARY CHECKLIST

Check your knowledge by attempting to write definitions of the terms below. Then compare your work with the definitions given in the text's Glossary. Ask your instructor which definitions, if any, might appear on a test.

associative law of multiplication: ....

block matrix: *See* partitioned matrix.

block matrix multiplication: The ... multiplication of ... as if ....

column sum: The sum of ....

commuting matrices: Two matrices A and B such that ....

composition of linear transformations: A mapping produced by applying ....

conformable for block multiplication: Two partitioned matrices A and B such that ....

consumption matrix: A matrix in the ... model whose columns are ....

diagonal entries (in a matrix): Entries having ....

diagonal matrix: A square matrix whose ... entries are ....

determinant of $A = \begin{bmatrix} a & b \\ c & d \end{bmatrix}$: The number ..., denoted by ....

distributive laws: (left) ... (right) ....

elementary matrix: An invertible matrix that results by ....

final demand vector (or **bill of final demands**): The vector **d** in the ... model that lists .... The vector **d** can represent ....

flexibility matrix: A matrix whose *j*th column gives ... of an elastic beam at specified points when ... is applied at ....

Gauss-Seidel algorithm: An iterative method that produces ... that in certain cases converges to a ...; based on the decomposition $A = M - N$, with M ....

Householder reflection: A transformation $\mathbf{x} \mapsto Q\mathbf{x}$, where $Q =$ ....

identity matrix: The $n \times n$ matrix I or I_n with ....

inner product: A matrix product ... where **u** and **v** are ....

input-output matrix: *See* consumption matrix.

input-output model: *See* Leontief input-output model.

intermediate demands: Demands for goods or services that ....

inverse (of an $n \times n$ matrix A): An $n \times n$ matrix A^{-1} such that ....

invertible linear transformation: A linear transformation $T: \mathbb{R}^n \to \mathbb{R}^n$ such that there exists ....

invertible matrix: A square matrix that ....

Jacobi's method: An iterative method that produces ... that in certain cases converges to a ...; based on the decomposition $A = M - N$, with M ....

ladder network: An electrical network assembled by connecting ....

left inverse (of A): Any rectangular matrix C such that ....

Leontief input-output model (or **Leontief production equation**): The equation ..., where ....

lower triangular matrix: A matrix with ....

lower triangular part (of A): A ... matrix whose entries on ....

LU factorization: The representation of a matrix A in the form $A = LU$, where L is ... and U is ....

main diagonal (of a matrix): The location of the ....

outer product: A matrix product ... where **u** and **v** are ....

partitioned matrix: A matrix whose entries are .... Sometimes called ....

permuted lower triangular matrix: A matrix such that ....

permuted LU factorization: The representation of a matrix A in the form $A = LU$ where L is ... and U is ....

production vector: The vector in the ... model that lists ....

right inverse (of A): Any rectangular matrix C such that ....

row-column rule: The rule for computing a product AB in which ....

Schur complement: A certain matrix formed from the blocks of a 2×2 partitioned matrix $A = [A_{ij}]$. If A_{11} is invertible, its Schur complement is given by .... If A_{22} is invertible, its Schur complement is given by ....

stiffness matrix: The inverse of a ... matrix. The jth column of a stiffness matrix gives ... at specified points on an elastic beam in order to produce ....

transfer matrix: A matrix A associated with an electrical circuit having input and output terminals, such that ....

transpose (of A): An $n \times m$ matrix A^T whose ... are the corresponding ... of ....

unit consumption vector: A column vector in the ... model that lists ....

unit lower triangular matrix: A ... matrix with ....

upper triangular matrix: A matrix U with ....

Vandermonde matrix: An $n \times n$ matrix V or its transpose, of the form ....

4

Determinants

4.1 INTRODUCTION TO DETERMINANTS

This section is relatively short and easy. Some exercises provide computational practice and others allow you to discover properties of determinants to be studied in the next section. You will enjoy Section 4.2 more if you finish your work on this section first.

KEY IDEAS

The long paragraph at the beginning of the section sets the stage for what follows. Read it quickly, without worrying about the details of the row operations. The main idea is that the determinant of A is a number that appears along the diagonal of an echelon form of A, and this number is nonzero if and only if the matrix is invertible. Later (on pages 170 and 180) you will understand why this idea is important. For now, this 3×3 case is only used to motivate the definition of $\det A$.

Determinants are defined here via a cofactor expansion along the first row. Since the cofactors involve determinants of smaller matrices, the definition is said to be *recursive*. For each $n \geq 2$, the determinant of an $n \times n$ matrix is based on the definition of the determinant of an $(n-1) \times (n-1)$ matrix. There are other equivalent definitions of the determinant, but we shall not digress to discuss them.

Study Tip: Watch how parentheses are used in Example 2 to avoid a common mistake. The cofactor expansion puts a minus sign in front of a_{32} because $(-1)^{3+2} = -1$. Since a_{32} happens to be negative, the correct term in the expansion is $-(-2)\det A_{32}$, *not* $-2 \det A_{32}$.

SOLUTIONS TO EXERCISES

1. By definition, $\det A$ is computed via a cofactor expansion along the first row:

$$\begin{vmatrix} 3 & 0 & 4 \\ 2 & 3 & 2 \\ 0 & 5 & -1 \end{vmatrix} = 3\begin{vmatrix} 3 & 2 \\ 5 & -1 \end{vmatrix} - 0\begin{vmatrix} 2 & 2 \\ 0 & -1 \end{vmatrix} + 4\begin{vmatrix} 2 & 3 \\ 0 & 5 \end{vmatrix}$$

$$= 3(-3 - 10) - 0 + 4(10 - 0) = -39 + 40 = 1$$

For comparison, a cofactor expansion down the second column yields

$$\begin{vmatrix} 3 & 0 & 4 \\ 2 & 3 & 2 \\ 0 & 5 & -1 \end{vmatrix} = (-1)^{1+2}\cdot 0\begin{vmatrix} 2 & 2 \\ 0 & -1 \end{vmatrix} + (-1)^{2+2}\cdot 3\begin{vmatrix} 3 & 4 \\ 0 & -1 \end{vmatrix} + (-1)^{3+2}\cdot 5\begin{vmatrix} 3 & 4 \\ 2 & 2 \end{vmatrix}$$

$$= 0 + 3(-3 - 0) - 5(6 - 8) = -9 + 10 = 1$$

Study Tip: To save time, omit the zero terms in a cofactor expansion.

7. By definition,

$$\begin{vmatrix} 4 & 3 & 0 \\ 6 & 5 & 2 \\ 9 & 7 & 3 \end{vmatrix} = 4\begin{vmatrix} 5 & 2 \\ 7 & 3 \end{vmatrix} - 3\begin{vmatrix} 6 & 2 \\ 9 & 3 \end{vmatrix} = 4(15 - 14) -3(18 - 18) = 4$$

Using the second column of A instead,

$$\begin{vmatrix} 4 & 3 & 0 \\ 6 & 5 & 2 \\ 9 & 7 & 3 \end{vmatrix} = -3\begin{vmatrix} 6 & 2 \\ 9 & 3 \end{vmatrix} + 5\begin{vmatrix} 4 & 0 \\ 9 & 3 \end{vmatrix} - 7\begin{vmatrix} 4 & 0 \\ 6 & 2 \end{vmatrix}$$

$$= -3(18 - 18) + 5(12 - 0) - 7(8 - 0) = 0 + 60 - 56 = 4$$

13. Row 2 or column 2 are the best choices because they contain the most zeros. We'll use row 2. Since the only nonzero entry in that row is 2, the determinant is $(-1)^{2+3}\cdot 2\cdot A_{23}$.

$$\det A = (-1)^{2+3}\cdot 2\cdot \begin{vmatrix} 4 & 0 & -1 & 3 & -5 \\ 0 & 0 & 2 & 0 & 0 \\ 7 & 3 & -6 & 4 & -8 \\ 5 & 0 & 5 & 2 & -3 \\ 0 & 0 & 9 & -1 & 2 \end{vmatrix} = (-2)\cdot \begin{vmatrix} 4 & 0 & 3 & -5 \\ 7 & 3 & 4 & -8 \\ 5 & 0 & 2 & -3 \\ 0 & 0 & -1 & 2 \end{vmatrix}$$

The best choice for this 4×4 determinant is to expand down the second column. Notice that the cofactor associated with the 3 in the (2,2)-position is the (2,2)-cofactor of the 4×4 matrix. The original location of the "3" in the 5×5 matrix is irrelevant.

$$\det A = (-2)\cdot(-1)^{2+2}(3)\cdot\begin{vmatrix} 4 & 3 & -5 \\ 5 & 2 & -3 \\ 0 & -1 & 2 \end{vmatrix}$$

Finally, use column 1 (although row 3 would work as well).

$$\det A = (-2)\cdot(-1)^{2+2}(3)\cdot\left\{4\begin{vmatrix} 2 & -3 \\ -1 & 2 \end{vmatrix} - 5\begin{vmatrix} 3 & -5 \\ -1 & 2 \end{vmatrix}\right\}$$

$$= -6\left[4(4 - 3) - 5(6 - 5)\right] = -6(4 - 5) = 6$$

Checkpoint: Try to complete the following statement: "If the kth column of the $n \times n$ identity matrix is replaced by a column vector $\mathbf{x}$ whose entries are $x_1, \ldots, x_n$, then the determinant of the resulting matrix is _____." To discover the answer, compute the determinants of the following matrices:

a. $\begin{bmatrix} 1 & 3 & 0 & 0 \\ 0 & 4 & 0 & 0 \\ 0 & 5 & 1 & 0 \\ 0 & 6 & 0 & 1 \end{bmatrix}$ b. $\begin{bmatrix} 1 & 0 & 3 & 0 \\ 0 & 1 & 4 & 0 \\ 0 & 0 & 5 & 0 \\ 0 & 0 & 6 & 1 \end{bmatrix}$ c. $\begin{bmatrix} 1 & 0 & 3 & 0 & 0 \\ 0 & 1 & 4 & 0 & 0 \\ 0 & 0 & 5 & 0 & 0 \\ 0 & 0 & 6 & 1 & 0 \\ 0 & 0 & 7 & 0 & 1 \end{bmatrix}$

19. $\det\begin{bmatrix} a & b \\ c & d \end{bmatrix} = ad - bc$, and $\det\begin{bmatrix} c & d \\ a & b \end{bmatrix} = cb - da = -\det\begin{bmatrix} a & b \\ c & d \end{bmatrix}$.

Interchanging two rows reverses the sign of the determinant, at least for the 2×2 case. Perhaps this is true for larger matrices.

25. The matrix is triangular, so use Theorem 2.

$$\det\begin{bmatrix} 1 & 0 & 0 \\ 0 & 1 & 0 \\ 0 & k & 1 \end{bmatrix} = 1\cdot1\cdot1 = 1 \quad \text{Product of the diagonal entries}$$

31. A 3×3 row replacement matrix has one of the following forms:

$$\begin{bmatrix} 1 & 0 & 0 \\ k & 1 & 0 \\ 0 & 0 & 1 \end{bmatrix}, \begin{bmatrix} 1 & 0 & 0 \\ 0 & 1 & 0 \\ k & 0 & 1 \end{bmatrix}, \begin{bmatrix} 1 & 0 & 0 \\ 0 & 1 & 0 \\ 0 & k & 1 \end{bmatrix},$$

$$\begin{bmatrix} 1 & k & 0 \\ 0 & 1 & 0 \\ 0 & 0 & 1 \end{bmatrix}, \begin{bmatrix} 1 & 0 & k \\ 0 & 1 & 0 \\ 0 & 0 & 1 \end{bmatrix}, \begin{bmatrix} 1 & 0 & 0 \\ 0 & 1 & k \\ 0 & 0 & 1 \end{bmatrix}$$

In each case the matrix is triangular with 1's on the diagonal, so its determinant equals 1. The determinant of a row replacement matrix is 1, at least for the 3×3 case. Perhaps this is true for larger matrices.

37. $\det A = \det\begin{bmatrix} 3 & 1 \\ 4 & 2 \end{bmatrix} = 3(2) - 1(4) = 6 - 4 = 2.$ Since $5A = \begin{bmatrix} 5\cdot3 & 5\cdot1 \\ 5\cdot4 & 5\cdot2 \end{bmatrix}$,

$$\det 5A = (5\cdot3)(5\cdot2) - (5\cdot1)(5\cdot4) = 150 - 100 = 50$$

So $\det 5A \neq 5\cdot\det A$. Can you see what the true relation between $\det 5A$ and $\det A$ really is, at least for this example? What about $\det 5A$ for any 2×2 matrix? Don't peek in Section 4.2. Try to work out the details for yourself. If you are clever, you might be able to guess (and perhaps verify) a formula for $\det rA$, where r is any scalar and A is any $n\times n$ matrix.

39. $\det[\mathbf{u}\ \ \mathbf{v}] = \det\begin{bmatrix} 3 & 1 \\ 0 & 2 \end{bmatrix} = 6$, $\det[\mathbf{u}\ \ \mathbf{x}] = \det\begin{bmatrix} 3 & x \\ 0 & 2 \end{bmatrix} = 6$, and the areas of the parallelograms determined by $[\mathbf{u}\ \ \mathbf{v}]$ and $[\mathbf{u}\ \ \mathbf{x}]$ both equal 6. To see why the areas are equal, consider the parallelograms determined by $\mathbf{u} = (3,0)$ and $\mathbf{v} = (1,2)$ and by $\mathbf{u}$ and $\mathbf{x} = (x,2)$:

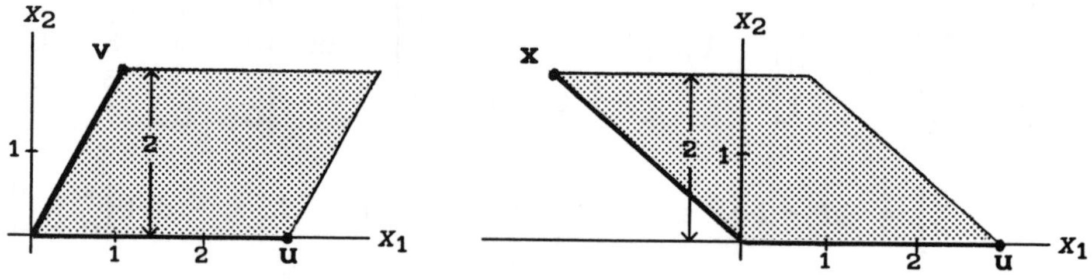

The parallelogram on the left is determined by $\mathbf{u}$ and $\mathbf{v}$ (and the vertices $\mathbf{u}+\mathbf{v}$ and $\mathbf{0}$). Its base is 3 and its altitude is 2, so the area is (base)(altitude) = 6. The parallelogram on the right, determined by $\mathbf{u}$ and $\mathbf{x} = (x,2)$, has the same base. Also, the altitude is 2 for any value of x, so the area again equals $3\cdot2 = 6$.

Answer to Checkpoint: a. 4 b. 5 c. 5 "If the kth column of the $n\times n$ identity matrix is replaced by a column vector $\mathbf{x}$ whose entries are $x_1,\ldots,x_n$, then the determinant of the resulting matrix is x_k." Can you explain why this is true? You'll learn the answer when you begin Section 4.3.

4.2 PROPERTIES OF DETERMINANTS

This section presents the main properties of determinants, gives an efficient method of computation, and proves that a matrix is invertible if and only if its determinant is nonzero.

KEY IDEAS

It is not surprising that row operations relate nicely to determinants. After all, we found the definition of a 3×3 determinant by row reducing a 3×3 matrix. Theorem 3 can be rephrased informally as follows:

a. *Adding a multiple of one row (or column) of* A *to another does not change the determinant.*

b. *Interchanging two rows (or columns) of* A *reverses the sign of the determinant.*

c. *A constant may be factored out of one row (or column) of the determinant of* A.

The other properties to learn are stated in Theorems 4, 5, and 6, together with the boxed formula for $\det A$ on page 170. Theorems 4 and 6 will be used extensively in Chapter 6.

Some texts say that a square matrix A is *singular* if its determinant is zero and *nonsingular* if the determinant is nonzero. Theorem 4 shows that the adjectives nonsingular and invertible are interchangeable.

Your instructor may or may not want you to know the (multi-) linearity property on pages 172-173. This property is important in more advanced courses but is not used later in the text. *Warning*: in general, $\det(A + B)$ is *unequal* to $\det A + \det B$.

SOLUTIONS TO EXERCISES

1. Rows 1 and 2 are interchanged, so the determinant changes sign.

7. $$\begin{vmatrix} 1 & 3 & 0 & 2 \\ -2 & -5 & 7 & 4 \\ 3 & 5 & 2 & 1 \\ 1 & -1 & 2 & -3 \end{vmatrix} = \begin{vmatrix} 1 & 3 & 0 & 2 \\ 0 & 1 & 7 & 8 \\ 0 & -4 & 2 & -5 \\ 0 & -4 & 2 & -5 \end{vmatrix} = \begin{vmatrix} 1 & 3 & 0 & 2 \\ 0 & 1 & 7 & 8 \\ 0 & 0 & 30 & 27 \\ 0 & 0 & 30 & 27 \end{vmatrix} = \begin{vmatrix} 1 & 3 & 0 & 2 \\ 0 & 1 & 7 & 8 \\ 0 & 0 & 30 & 27 \\ 0 & 0 & 0 & 0 \end{vmatrix} = 0$$

Note, the second array already shows that the determinant is zero, because two rows are equal, as in Example 3.

Study Tip: In general, computation of a 3×3 determinant by row reduction takes 10 multiplications (and divisions), but cofactor expansion only takes

9 multiplications. At $n = 4$, the advantage switches to row reduction, which requires 23 multiplications, cofactor expansion 40 (9 for each 3×3 determinant, plus four multiplications of a_{ij} times $\det A_{ij}$). Often, the best strategy is to combine the two techniques, as in Exercises 11-14.

13. Use row or column operations whenever convenient to create a row or column that has only one nonzero entry. (I recommend using only row operations, because you already have experience with them.) Then use a cofactor expansion to reduce the size of the matrix.

$$\begin{vmatrix} 2 & 5 & 4 & 1 \\ 4 & 7 & 6 & 2 \\ 6 & -2 & -4 & 0 \\ -6 & 7 & 7 & 0 \end{vmatrix} = \begin{vmatrix} 2 & 5 & 4 & 1 \\ 0 & -3 & -2 & 0 \\ 6 & -2 & -4 & 0 \\ -6 & 7 & 7 & 0 \end{vmatrix}$$ Zero created in column 4

$$= -\begin{vmatrix} 0 & -3 & -2 \\ 6 & -2 & -4 \\ -6 & 7 & 7 \end{vmatrix}$$ Result of cofactor expansion down column 4

$$= -\begin{vmatrix} 0 & -3 & -2 \\ 6 & -2 & -4 \\ 0 & 5 & 3 \end{vmatrix}$$ Zero created in column 1

$$= -(-6)\begin{vmatrix} -3 & -2 \\ 5 & 3 \end{vmatrix}$$ Result of cofactor expansion down column 1

$$= 6\cdot(-9 + 10) = 6$$

19. $$\begin{vmatrix} a & b & c \\ 2d+a & 2e+b & 2f+c \\ g & h & i \end{vmatrix} = \begin{vmatrix} a & b & c \\ 2d & 2e & 2f \\ g & h & i \end{vmatrix}$$ (-1)·row 1 added to row 2

$$= 2\begin{vmatrix} a & b & c \\ d & e & f \\ g & h & i \end{vmatrix}$$ 2 factored out of row 2

$$= 2\cdot 7 = 14$$

25. By Theorem 4 and the IMT, the set $\{\mathbf{v}_1, \mathbf{v}_2, \mathbf{v}_3\}$ is linearly independent if and only if $\det[\mathbf{v}_1 \ \ \mathbf{v}_2 \ \ \mathbf{v}_3] \neq 0$. Rather than use row operations on $[\mathbf{v}_1 \ \ \mathbf{v}_2 \ \ \mathbf{v}_3]$, you might choose to expand the determinant by cofactors of the third column:

$$\begin{vmatrix} 7 & -8 & 7 \\ -4 & 5 & 0 \\ -6 & 7 & -5 \end{vmatrix} = 7\begin{vmatrix} -4 & 5 \\ -6 & 7 \end{vmatrix} + (-5)\begin{vmatrix} 7 & -8 \\ -4 & 5 \end{vmatrix} = 7(-28 + 30) -5(35 - 32)$$

$$= 7(2) -5(3) = -1$$

The determinant is nonzero, so the vectors are linearly independent.

Study Tip: For 3×3 matrices, some students tend to prefer the special trick suggested for Exercises 15–18 in Section 4.1, even though in general there are 12 multiplications instead of the 9 multiplications needed for cofactor expansion. Note, however, that numbers in the special method can sometimes be large. For comparison, here are those computations for the matrix studied above in Exercise 25:

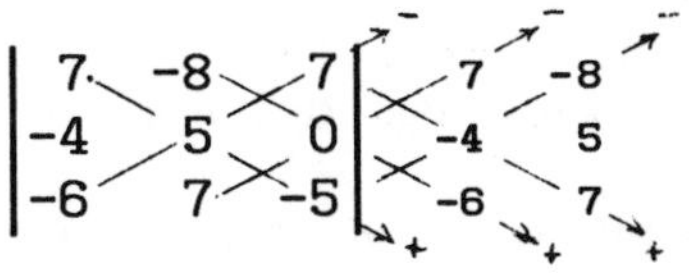

$$\begin{vmatrix} 7 & -8 & 7 \\ -4 & 5 & 0 \\ -6 & 7 & -5 \end{vmatrix} \begin{matrix} 7 & -8 \\ -4 & 5 \\ -6 & 7 \end{matrix}$$

$$\det \begin{bmatrix} \mathbf{v_1} & \mathbf{v_2} & \mathbf{v_3} \end{bmatrix} = 7(5)(-5) + (-8)(0)(-6) + 7(-4)(7)$$
$$-(-6)(5)(7) - 7(0)(7) - (-5)(-4)(-8)$$

$$= -175 + 0 + (-196) - (-210) - 0 -(-160) = -1$$

29. Since the determinant is multiplicative (Theorem 6),

$$(\det A)(\det A^{-1}) = \det(AA^{-1}) = \det I = 1. \text{ So } \det A^{-1} = 1/\det A.$$

Study Tip: The result of Exercise 29 might come in handy on a test.

31. By Theorem 6 (twice), $\det AB = (\det A)(\det B) = (\det B)(\det A) = \det BA$.

33. By Theorem 5, $\det U^T = \det U$. So, by Theorem 6,

$$\det U^T U = (\det U^T)(\det U) = (\det U)^2$$

If $U^T U = I$, then $(\det U)^2 = \det I = 1$, which implies that $\det U = \pm 1$.

37. Assume that $\det A = 4$ and $\det B = -3$.

a. $\det AB = (\det A)(\det B) = 4(-3) = -12$ Theorem 6

b. $\det 5A = \det \begin{bmatrix} 5a_{11} & 5a_{12} & 5a_{13} \\ 5a_{21} & 5a_{22} & 5a_{23} \\ 5a_{31} & 5a_{32} & 5a_{33} \end{bmatrix} = 5^3 \det A$ Theorem 3(c) three times

$$= 5^3 \cdot 4 = 500$$

c. $\det B^T = \det B = -3$ Theorem 5

d. $\det A^{-1} = 1/\det A = 1/4$ — Exercise 29

e. $\det A^3 = \det A \cdot \det A^2 = (\det A)^3 = (4)^3 = 64$ — Theorem 6 two times

Study Tip: Exercises 15-30, 37 and 38 make good test questions because they check your knowledge of determinant properties without requiring much computation. Exercise 37(b) is the one most likely to be answered incorrectly. What would be the answer to 37(b) if A were 4×4?

41. Compute det A by a cofactor expansion down column 3:

$$\det A = (u_1 + v_1)\cdot\det A_{13} - (u_2 + v_2)\cdot\det A_{23} + (u_3 + v_3)\cdot\det A_{33}$$

$$= u_1\cdot\det A_{13} - u_2\cdot\det A_{23} + u_3\cdot\det A_{33} + v_1\cdot\det A_{13} - v_2\cdot\det A_{23} + v_3\cdot\det A_{33}$$

$$= u_1\cdot\det B_{13} - u_2\cdot\det B_{23} + u_3\cdot\det B_{33} + v_1\cdot\det C_{13} - v_2\cdot\det C_{23} + v_3\cdot\det C_{33}$$

$$= \det B + \det C$$

MATLAB To compute det A, set **U = A** and then repeatedly use the commands **U = gauss(U,r)** and **U = swap(U,r,s)** as needed to reduce A to an echelon form U. (See the MATLAB notes for Section 3.2.) Keep track of how many times you swap rows. Then, except for a ±1, the determinant of A is given by the command

```
prod(diag(U))
```

The command **diag(U)** extracts the diagonal entries of U and places them in a column vector, and **prod** computes the product of those entries. You can, of course, simply use **det(A)** to check your work, but the longer sequence of commands helps you think about the *process* of computing det A.

4.3 CRAMER'S RULE, VOLUME, AND LINEAR TRANSFORMATIONS

This section will be a valuable reference for students who plan to take a course in multivariable calculus. Mathematics and statistics majors probably will encounter the material here several times. Also, economics students and engineers (particularly electrical engineers) are likely to need Cramer's rule and some of the supplementary exercises in later courses.

KEY IDEAS

The main results of the section are stated in Theorems 7, 8, 9, and 10. The proof of Theorem 7 is simple and yet involves three important ideas: the definition of a matrix product, the multiplicative property of the determinant, and the evaluation of a determinant by cofactors. Check with your instructor about whether you should be able to prove Theorem 7.

Cramer's Rule and the adjugate of a matrix are primarily theoretical tools, though sometimes useful for 2×2 and 3×3 matrices. For computations with larger matrices, you should avoid determinants!

A heuristic proof of Theorem 9 for 2×2 matrices is given in an appendix at the end of this section. Theorem 10 provides a key idea in calculus and physics needed for the study of double and triple integrals. The matrix used there is called a *Jacobian*.

STUDY NOTE

In Exercise 25, you are asked to use Theorem 9 to explain why the determinant of a 3×3 matrix A is zero if and only if A is not invertible. (A similar explanation holds for the 2×2 case.) The answer is in the text, so be sure to work on this by yourself before looking at the answer section. *Work on Exercise 25, even if it is not assigned.*

Remember, learning *does* take place when you think hard about an exercise, even when you are unsuccessful, if you try to look at the problem from different angles, browse back through the text, and perhaps look at earlier exercises. *Write* your solution, don't just talk to yourself about what you would write if you had to. Allow yourself at least 10 minutes, if necessary, to work on Exercise 25 before turning to the answer section.

SOLUTIONS TO EXERCISES

1. The system is equivalent to $A\mathbf{x} = \mathbf{b}$, where $A = \begin{bmatrix} 5 & 7 \\ 2 & 4 \end{bmatrix}$ and $\mathbf{b} = \begin{bmatrix} 3 \\ 1 \end{bmatrix}$.

   Write

   $$A_1(\mathbf{b}) = \begin{bmatrix} 3 & 7 \\ 1 & 4 \end{bmatrix}, \quad A_2(\mathbf{b}) = \begin{bmatrix} 5 & 3 \\ 2 & 1 \end{bmatrix}$$
   $$\qquad\qquad \uparrow \qquad\qquad\qquad\qquad \uparrow$$
   $$\qquad\qquad \mathbf{b} \qquad\qquad\qquad\qquad \mathbf{b}$$

   and compute

   $$\det A = 20 - 14 = 6, \quad \det A_1(\mathbf{b}) = 12 - 7 = 5, \quad \det A_2(\mathbf{b}) = 5 - 6 = -1$$

   $$x_1 = \frac{\det A_1(\mathbf{b})}{\det A} = \frac{5}{6}, \quad x_2 = \frac{\det A_2(\mathbf{b})}{\det A} = \frac{-1}{6} = -\frac{1}{6}$$

7. The system is equivalent to $A\mathbf{x} = \mathbf{b}$, where $A = \begin{bmatrix} 6s & 4 \\ 9 & 2s \end{bmatrix}$ and $\mathbf{b} = \begin{bmatrix} 5 \\ -2 \end{bmatrix}$. Write

$$A_1(\mathbf{b}) = \begin{bmatrix} 5 & 4 \\ -2 & 2s \end{bmatrix}, \quad A_2(\mathbf{b}) = \begin{bmatrix} 6s & 5 \\ 9 & -2 \end{bmatrix}$$

and compute

$$\det A = 12s^2 - 36 = 12(s^2 - 3) = 12(s - \sqrt{3})(s + \sqrt{3})$$

$$\det A_1(\mathbf{b}) = 10s + 8, \quad \det A_2(\mathbf{b}) = -12s - 45$$

The system has a unique solution when $\det A \neq 0$, that is, when $s \neq \pm\sqrt{3}$. For such a system, the solution is $\mathbf{x} = (x_1, x_2)$, where

$$x_1 = \frac{\det A_1(\mathbf{b})}{\det A} = \frac{10s + 8}{12(s^2 - 3)} = \frac{5s + 4}{6(s^2 - 3)}$$

$$x_2 = \frac{\det A_2(\mathbf{b})}{\det A} = \frac{-12s - 45}{12(s^2 - 3)} = \frac{-4s - 15}{4(s^2 - 3)}$$

13. First, find the cofactors of $A = \begin{bmatrix} 3 & 5 & 4 \\ 1 & 0 & 1 \\ 2 & 1 & 1 \end{bmatrix}$.

$$C_{11} = +\begin{vmatrix} 0 & 1 \\ 1 & 1 \end{vmatrix} = -1, \quad C_{12} = -\begin{vmatrix} 1 & 1 \\ 2 & 1 \end{vmatrix} = 1, \quad C_{13} = +\begin{vmatrix} 1 & 0 \\ 2 & 1 \end{vmatrix} = 1$$

$$C_{21} = -\begin{vmatrix} 5 & 4 \\ 1 & 1 \end{vmatrix} = -1, \quad C_{22} = +\begin{vmatrix} 3 & 4 \\ 2 & 1 \end{vmatrix} = -5, \quad C_{23} = -\begin{vmatrix} 3 & 5 \\ 2 & 1 \end{vmatrix} = 7$$

$$C_{31} = +\begin{vmatrix} 5 & 4 \\ 0 & 1 \end{vmatrix} = 5, \quad C_{32} = -\begin{vmatrix} 3 & 4 \\ 1 & 1 \end{vmatrix} = 1, \quad C_{33} = +\begin{vmatrix} 3 & 5 \\ 1 & 0 \end{vmatrix} = -5$$

Then, arrange the *transpose* of the array of cofactors into the adjugate of A.

$$\operatorname{adj} A = \begin{bmatrix} -1 & -1 & 5 \\ 1 & -5 & 1 \\ 1 & 7 & -5 \end{bmatrix}$$

Were you to compute $\det A$ now, you could write A^{-1}, but you would still need to check whether your calculations are correct. To build in this check, compute

$$A \cdot \text{adj}\, A = \begin{bmatrix} 3 & 5 & 4 \\ 1 & 0 & 1 \\ 2 & 1 & 1 \end{bmatrix} \begin{bmatrix} -1 & -1 & 5 \\ 1 & -5 & 1 \\ 1 & 7 & -5 \end{bmatrix} = \begin{bmatrix} 6 & 0 & 0 \\ 0 & 6 & 0 \\ 0 & 0 & 6 \end{bmatrix}$$

If any off-diagonal entries in the product are nonzero, or if the diagonal entries are not all the same, then some errors have been made, and you can recheck your cofactor calculations. (One possible mistake is to forget the ± signs in front of the 2×2 determinants. Another error is to *not* transpose the array of cofactors.) In this case, the calculations above *are* correct and det A must be 6. So

$$A^{-1} = \frac{1}{\det A} \text{adj}\, A = \frac{1}{6} \begin{bmatrix} -1 & -1 & 5 \\ 1 & -5 & 1 \\ 1 & 7 & -5 \end{bmatrix}$$

19. The parallelogram with vertices (0,0), (5,2), (6,4), (11,6) is shown below. If no vertex were zero, we would have to translate the parallelogram to the origin by subtracting one vertex from all four vertices. Since one vertex already is zero, use the two vertices adjacent to the origin to construct the columns of A, and compute $|\det A|$.

$$A = \begin{bmatrix} 5 & 6 \\ 2 & 4 \end{bmatrix}, \quad \left\{\begin{matrix} \text{area of the} \\ \text{parallelogram} \end{matrix}\right\} = |\det A| = |20 - 12| = 8$$

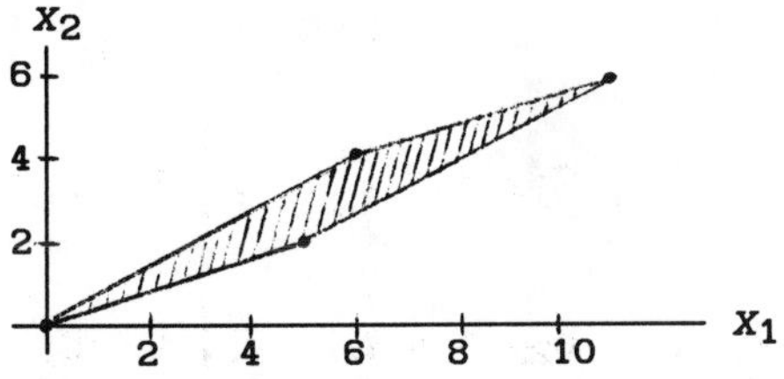

25. The answer is in the text. I hope you took the advice at the beginning of this Study Guide section and worked the problem (or at least tried hard to work the problem) before checking the answer section. If you were successful, you should be proud of yourself; you are mastering the material—not only determinants but also linear dependence!

I'm sorry to say that if you looked at the answer before really trying to create the solution, you have ruined a good problem. Once you have read the answer, you can never go back, recapture your innocence, and have a chance to feel the rush of success. But you will have other chances later in the text, with other good exercises!

31. Let $\mathbf{x} = \begin{bmatrix} x_1 \\ x_2 \\ x_3 \end{bmatrix}$, $\mathbf{u} = \begin{bmatrix} u_1 \\ u_2 \\ u_3 \end{bmatrix}$, and $A = \begin{bmatrix} a & 0 & 0 \\ 0 & b & 0 \\ 0 & 0 & c \end{bmatrix}$. Also, let S be the unit

   ball in $\mathbb{R}^3$, whose bounding surface consists of all vectors $\mathbf{u}$ such that $u_1^2 + u_2^2 + u_3^2 = 1$, and let S' be the image of S under the mapping $\mathbf{u} \mapsto A\mathbf{u}$.

   a. If $\mathbf{x}$ is in S', then $\mathbf{x} = A\mathbf{u}$ for some $\mathbf{u}$ in S, and $\mathbf{u} = A^{-1}\mathbf{x} = \begin{bmatrix} x_1/a \\ x_2/b \\ x_3/c \end{bmatrix}$.

   The condition on u_1, u_2, u_3 shows that $\left(\dfrac{x_1}{a}\right)^2 + \left(\dfrac{x_2}{b}\right)^2 + \left(\dfrac{x_3}{c}\right)^2 = 1$.

   b. Since the volume of the unit ball bounded by S is $4\pi/3$ and the determinant of A is abc, Theorem 10 shows that the volume of the region bounded by S' is $4\pi abc/3$.

Appendix: A Geometric Proof of a Determinant Property

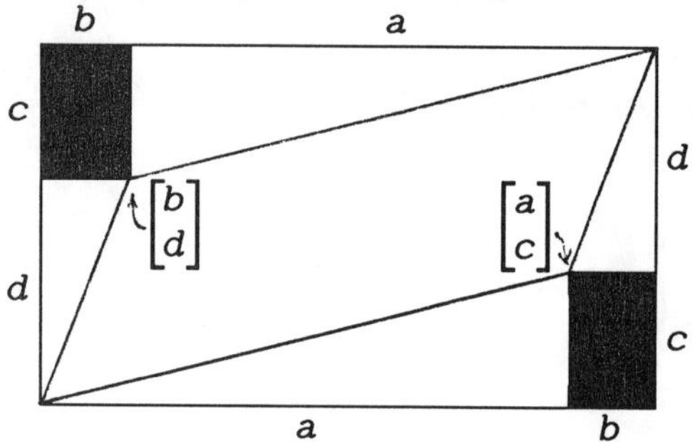

$$\det \begin{bmatrix} a & b \\ c & d \end{bmatrix} = ad - bc$$

$$(a + b)(c + d) = ac + ad + bc + bd$$

$$- 2 \begin{bmatrix} & c \\ b & \end{bmatrix} - 2 \begin{bmatrix} & d \\ b & \end{bmatrix} = - 2bc - bd$$

$$- 2 \begin{bmatrix} & c \\ a & \end{bmatrix} = - ac$$

(Area of the Parallelogram)   = $ad - bc$

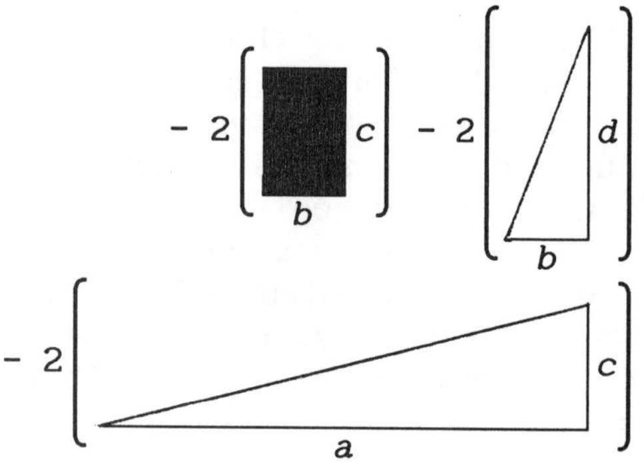

CHAPTER 4 GLOSSARY CHECKLIST

Check your knowledge by attempting to write definitions of the terms below. Then compare your work with the definitions given in the text's Glossary. Ask your instructor which definitions, if any, might appear on a test.

adjugate (or **classical adjoint**): The matrix adj A formed from a square matrix A by replacing the (i, j)-entry of A by ....

cofactor: A number C_{ij} = ..., called the (i, j)-*cofactor of* A, where A_{ij} is the submatrix formed by ....

cofactor expansion: A formula for det A using cofactors associated with one row or one column, such as for row 1: ....

Cramer's Rule: A formula for ....

5

Vector Spaces

5.1 VECTOR SPACES AND SUBSPACES

The main focus of the chapter is on $\mathbb{R}^n$ and its subspaces. However, Section 5.1 builds a framework within which the theory for $\mathbb{R}^n$ rests. Most of the exercises in the chapter concern subspaces of $\mathbb{R}^n$, but some are designed to help you learn gradually about other important vector spaces.

KEY IDEAS

A *vector space* is any collection of objects that behave as vectors do in $\mathbb{R}^n$. (The precise meaning of "behave" is described by the axioms on page 191.) A *vector* is simply any object that belongs to a vector space. Lists of numbers, arrows, and polynomials are all examples of vectors, in different vector spaces.

The most important vector spaces in this text are subspaces of $\mathbb{R}^n$, which are visualized as lines and planes *through the origin*, the origin by itself, and the entire space $\mathbb{R}^n$. To show that a set is a vector space, use Theorem 1 or 2. (Theorem 2 is easier to use, if it applies.) To show that a set is *not* a subspace, use Theorem 1. (See Exercises 1-4.)

STUDY NOTES

Parts of this chapter are somewhat more theoretical than the earlier chapters, but that is necessary in order to give a solid foundation for the rest of the course. The advice given earlier in this Study Guide is even more important now: it is not enough to read the section once and then to work the assigned homework problems. You need to thoroughly learn the key definitions and theorems. The writing exercises will help you do this. You cannot wait until just before an exam to learn theorems, because you will need to know them for use in the exercises and in subsequent sections.

In Example 5, the concept of a function as a single "vector" in a vector space is difficult to absorb on a first reading, and you should not expect to master it in a few days.

The word *subspace* should usually be accompanied by a phrase such as "*of* V", or "*of* $\mathbb{R}^n$". In each case, the symbol or phrase following "*of*" determines the general nature of the elements in the subspace; this symbol or phrase does *not* give the specific condition(s) that define the subspace. For instance, the statement, "H is a subspace of $\mathbb{P}_3$" means that each vector in H is a polynomial of degree at most 3. More information about H would be necessary in order to determine *which* subspace of $\mathbb{P}_3$ is in view.

Set Notation: The notation introduced in Example 8 will be used frequently in this chapter as an efficient way to describe a set. The set description has two parts: the first part, to the left of the colon, shows or describes the basic nature of the elements of the set; the second part is a statement that adds any qualifying conditions that must be met in order for an element to belong to the set. There are usually several ways to describe a given set. For instance, the set in Example 8 could also be written as

$$H = \left\{ \begin{bmatrix} s \\ t \\ u \end{bmatrix} : s \text{ and } t \text{ are real, and } u = 0 \right\}$$

As another example, the set in Exercise 11 could be written as

$$W = \left\{ \begin{bmatrix} a \\ b \\ c \end{bmatrix} : a = 5b + 2c, \text{ where } b \text{ and } c \text{ are arbitrary} \right\}$$

Here, and elsewhere, reference to the scalars b and c as "arbitrary" means that the scalars can be any real numbers.

Study Tip: Review Sections 2.1 to 2.4 before you reach Section 5.3.

SOLUTIONS TO EXERCISES

1. a. V is a subset of $\mathbb{R}^2$. The defining property of V is that the entries of every vector in V are nonnegative. So, if **u** and **v** are in V, their entries are nonnegative. Since the entries in **u** + **v** are also nonnegative, the vector **u** + **v** satisfies the condition that defines V. That is, **u** + **v** is in V.

   b. The text gives a specific **u** and c. You could also simply say, "If a nonzero vector **v** in V is multiplied by a negative scalar c, then c**v** is not in V because at least one of its entries is negative." If such a general statement is hard to compose, try to find a specific "counterexample" to one part of the Subspace Test.

Study Tip: Remember that in order to show that a given set W of vectors is **not** a subspace, it suffices to find one specific example where one of the three conditions of the subspace test is *not* satisfied. Check the zero vector first.

7. Most examples in the text involve integers, to make calculations simple, and one can easily overlook the fact that a subspace must be closed under multiplication by *all* real numbers.

13. a. $\mathbf{w}$ is certainly not one of the three vectors in $\{\mathbf{v}_1, \mathbf{v}_2, \mathbf{v}_3\}$.

b. Span$\{\mathbf{v}_1, \mathbf{v}_2, \mathbf{v}_3\}$ contains infinitely many vectors.

c. $\mathbf{w}$ is in the subspace (of $\mathbb{R}^3$) spanned by $\mathbf{v}_1, \mathbf{v}_2, \mathbf{v}_3$ if and only if the equation $x_1\mathbf{v}_1 + x_2\mathbf{v}_2 + x_3\mathbf{v}_3 = \mathbf{w}$ is consistent (has a solution). Row reduce the augmented matrix:

$$\begin{bmatrix} 1 & 2 & 4 & 3 \\ 0 & 1 & 2 & 1 \\ -1 & 3 & 6 & 2 \end{bmatrix} \sim \begin{bmatrix} 1 & 2 & 4 & 3 \\ 0 & 1 & 2 & 1 \\ 0 & 5 & 10 & 5 \end{bmatrix} \sim \begin{bmatrix} 1 & 2 & 4 & 3 \\ 0 & 1 & 2 & 1 \\ 0 & 0 & 0 & 0 \end{bmatrix}$$

There is no pivot in the augmented column, so the vector equation is consistent, and $\mathbf{w}$ is in Span$\{\mathbf{v}_1, \mathbf{v}_2, \mathbf{v}_3\}$.

19. Let H be the set defined by (5). Then H is a subset of the vector space V of all real-valued functions, and H consists of all linear combinations of the functions $\cos \omega t$ and $\sin \omega t$. By Theorem 2, H is a subspace of V, and hence is a vector space.

25. The answers are in the text. Exercises 23-28 show how facts that are perfectly "obvious" in $\mathbb{R}^n$ depend only on a few basic properties of $\mathbb{R}^n$ that are now axioms for a general vector space.

30. (Comments, not the solution) If H and K are subspaces of a vector space V, then $H \cap K$ is the largest subspace of V that is contained in both H and K. A related subspace, not mentioned in the text, is the smallest subspace of V that contains both H and K. This subspace is denoted by $H + K$, and it consists of all vectors $\mathbf{v}$ in V that can be written in the form $\mathbf{v} = \mathbf{h} + \mathbf{k}$, where $\mathbf{h}$ is some vector in H and $\mathbf{k}$ is some vector in K. In set notation,

$$H + K = \{\mathbf{h} + \mathbf{k} : \mathbf{h} \text{ is in } H \text{ and } \mathbf{k} \text{ is in } K\}$$

If Exercise 30 is assigned as homework, there is a chance that somewhere along the way you might encounter the subspace $H + K$. (Guess where!)

5.2 NULL SPACES, COLUMN SPACES, AND LINEAR TRANSFORMATIONS

Many important concepts of linear algebra involve a subspace in one way or another. This section provides an opportunity to become familiar with the most important classes of subspaces. Most students need to think about subspaces awhile before they are comfortable with the concept. The foundation for this section was laid in Sections 2.1 and 2.3. Have you reviewed those sections yet?

KEY IDEAS

The proof of Theorem 3 is important. It makes a good exam question, because it tests to see if you understand both the Subspace Test and the definition of Nul A. If you know the statement of the Subspace Test and the definition of Nul A, the proof of Theorem 3 should be easy to remember.

Theorem 4 actually has two conclusions. The statement "Col A is a subspace" tells us that linear combinations of vectors in Col A remain in Col A. The phrase "of $\mathbb{R}^m$" reminds us that each vector has m entries (because A has m rows). Similarly, the phrase "of R^n" in Theorem 3 reminds us each vector in Nul A has n entries, in order for the product $A\mathbf{x}$ to be defined.

The boxed statement after Theorem 4 shows that the statement "Col $A = \mathbb{R}^m$" can be added to the list of equivalent statements in Theorem 2 of Section 2.2.

STUDY NOTES

In Example 4, the statement "Nul $A = \text{Span}\{\mathbf{u}, \mathbf{v}, \mathbf{w}\}$" would be an *explicit* description of Nul A (provided you specify what $\mathbf{u}$, $\mathbf{v}$, and $\mathbf{w}$ are).

In some applications, it is important to know that for a given $m \times n$ matrix A, every equation $A\mathbf{x} = \mathbf{b}$ has a solution (assuming $\mathbf{b}$ is in $\mathbb{R}^m$). Yet it may require some effort to determine whether this is the case. If not every equation $A\mathbf{x} = \mathbf{b}$ has a solution, then not every $\mathbf{b}$ belongs to Col A, and hence Col A is a proper subspace of $\mathbb{R}^m$. One of the goals of the next few sections is to obtain a method for determining when Col $A = \mathbb{R}^m$.

Checkpoint 1: How many pivot positions does an $m \times n$ matrix A have if Col $A = \mathbb{R}^m$?

Study Tip: Theorems 2, 3, and 4 are the main weapons for showing that a set *is* a vector space (that is, a subspace of some known vector space). Review these theorems now, before starting the exercises. The Subspace Test is the main tool for showing why some set **is not** a vector space.

SOLUTIONS TO EXERCISES

1. Now is the time to learn the *definition* of Nul A. A vector $\mathbf{x}$ is in Nul A precisely when the product $A\mathbf{x}$ is defined and $A\mathbf{x} = \mathbf{0}$. Given $\mathbf{x}$, simply compute $A\mathbf{x}$ to determine whether $A\mathbf{x}$ is zero.

$$A\mathbf{x} = \begin{bmatrix} 3 & -5 & -3 \\ 6 & -2 & 0 \\ -8 & 4 & 1 \end{bmatrix}\begin{bmatrix} 1 \\ 3 \\ -4 \end{bmatrix} = \begin{bmatrix} 0 \\ 0 \\ 0 \end{bmatrix}, \quad \text{so} \quad \begin{bmatrix} 1 \\ 3 \\ -4 \end{bmatrix} \text{ is in Nul } A.$$

Warning: In Exercises 3-6, writing an equation $\mathbf{x} = c\mathbf{u} + d\mathbf{v}$ is not the same as listing the vectors $\mathbf{u}$ and $\mathbf{v}$, say, that span the null space. The appropriate answer for these exercises is a list of a small finite number of vectors that span Nul A, not a description of *all* the vectors in Nul A.

Study Tip: Try Practice Problem 1 before you work on Exercises 7-18. If you can't find *two* ways to work the practice problem, reread the first paragraph of Section 5.2 (but don't look at it until you have attempted the practice problem).

7. The set W is a sub*set* of $\mathbb{R}^3$. If W were a vector space (under the standard operations in $\mathbb{R}^3$), it would be a sub*space* of $\mathbb{R}^3$. But W fails *every* test in the Subspace Test, so it is not a vector space. For instance, the vector (0,0,0) does not satisfy the condition $a + b + c = 2$, and so the zero vector is not in W.

13. A typical element of W can be written as follows:

$$\begin{bmatrix} a \\ b \\ 2a-b \end{bmatrix} = \begin{bmatrix} a \\ 0 \\ 2a \end{bmatrix} + \begin{bmatrix} 0 \\ b \\ -b \end{bmatrix} = a\underset{\uparrow\atop\mathbf{u}}{\begin{bmatrix} 1 \\ 0 \\ 2 \end{bmatrix}} + b\underset{\uparrow\atop\mathbf{v}}{\begin{bmatrix} 0 \\ 1 \\ -1 \end{bmatrix}}$$

This calculation shows that W coincides with Span$\{\mathbf{u},\mathbf{v}\}$, so W is a subspace of $\mathbb{R}^3$, by Theorem 2.

Checkpoint 2: Why is W a subspace "of $\mathbb{R}^3$"?

19. Write a typical element of W as

$$\begin{bmatrix} 2s+3t \\ r+s-2t \\ 4r+s \\ 3r-s-t \end{bmatrix} = r\begin{bmatrix} 0 \\ 1 \\ 4 \\ 3 \end{bmatrix} + s\begin{bmatrix} 2 \\ 1 \\ 1 \\ -1 \end{bmatrix} + t\begin{bmatrix} 3 \\ -2 \\ 0 \\ -1 \end{bmatrix} = A\begin{bmatrix} r \\ s \\ t \end{bmatrix}, \quad \text{where } A = \begin{bmatrix} 0 & 2 & 3 \\ 1 & 1 & -2 \\ 4 & 1 & 0 \\ 3 & -1 & -1 \end{bmatrix}$$

This calculation shows that W = Col A.

Study Tip: Exercises 21-24 may seem simple, but they will help you avoid a misunderstanding later in Section 5.5.

25. $A = \begin{bmatrix} 2 & -6 \\ -1 & 3 \\ -4 & 12 \\ 3 & -9 \end{bmatrix}$. Either $\begin{bmatrix} 2 \\ -1 \\ -4 \\ 3 \end{bmatrix}$ or $\begin{bmatrix} -6 \\ 3 \\ 12 \\ -9 \end{bmatrix}$ is an obvious choice for a vector in Col A. To find a vector in Nul A, find one solution of $A\mathbf{x} = \mathbf{0}$. So row reduce the augmented matrix $[A \ \ \mathbf{0}]$:

$$\begin{bmatrix} 2 & -6 & 0 \\ -1 & 3 & 0 \\ -4 & 12 & 0 \\ 3 & -9 & 0 \end{bmatrix} \sim \begin{bmatrix} 1 & -3 & 0 \\ 0 & 0 & 0 \\ 0 & 0 & 0 \\ 0 & 0 & 0 \end{bmatrix}, \quad \begin{matrix} x_1 - 3x_2 = 0, & x_1 = 3x_2 \\ x_2 \text{ is free} & \end{matrix}$$

The general solution of $A\mathbf{x} = \mathbf{0}$ is not needed, so choose any nonzero value for x_2, say $x_2 = 1$. Then $x_1 = 3$, and a vector in Nul A is $\begin{bmatrix} 3 \\ 1 \end{bmatrix}$.

Checkpoint 3: Is $\begin{bmatrix} 3 \\ 1 \\ 0 \\ 0 \end{bmatrix}$ also in Nul A? Why or why not?

31. The solutions are in the text. Make a note in your book to practice writing the solutions again before the next test, without looking at the back of the book. Make sure you write complete English sentences that *explain* what you are doing. In part (b), for example, the equation "$A\mathbf{x} + A\mathbf{w} = A(\mathbf{x} + \mathbf{w})$" by itself is *not* a satisfactory answer.

Important: Don't just "talk" to yourself about what you would write on a test, and then check to see if you said things correctly. Doing this might give you a false estimate of your mastery of this material.

37. Use the Subspace Test. First, $\mathbf{0}_W$ is in $T(U)$, because the linear transformation T takes the zero vector in V (which is also in U) onto the zero vector in W. Second, typical elements of $T(U)$ have the form $T(\mathbf{u}_1)$ and $T(\mathbf{u}_2)$, where $\mathbf{u}_1, \mathbf{u}_2$ are in U. Since T is linear,

$$T(\mathbf{u}_1) + T(\mathbf{u}_2) = T(\mathbf{u}_1 + \mathbf{u}_2). \qquad (*)$$

Since U is a subspace, $\mathbf{u}_1 + \mathbf{u}_2$ is in U, so the left side of (*) is in $T(U)$. Thus $T(U)$ is closed under vector addition. Third, $c \cdot T(\mathbf{u}_1) = T(c\mathbf{u}_1)$, for any scalar c, because T is linear. Since U is a subspace, $c\mathbf{u}_1$ is in U, and hence $c \cdot T(\mathbf{u}_1)$ is in $T(U)$. These three steps show that $T(U)$ is a subspace of W, by the Subspace Test.

Answers to Checkpoints:

1. An $m \times n$ matrix A must have m pivot positions in order for Col A to be all of $\mathbb{R}^m$.

2. W is a subspace " of $\mathbb{R}^3$ " because each vector in W has three entries.

3. No, the listed vector is not in $\mathbb{R}^2$. Since A has only two columns, vectors in Nul A must have only two entries. Otherwise, the product $A\mathbf{x}$ is not even defined, let alone equal to $\mathbf{0}$. (If you missed this question, you have not yet learned the definition of Nul A. Learn it now.)

5.3 LINEARLY INDEPENDENT SETS; BASES

The definition of linear independence carries over from $\mathbb{R}^n$ to any vector space. And the geometric interpretations in Chapter 2 of linearly independent and dependent sets should help you visualize these concepts here.

KEY IDEAS

In general, you cannot use an ordinary matrix equation $A\mathbf{x} = \mathbf{0}$ to study linear dependence of $\{\mathbf{v}_1, \ldots, \mathbf{v}_p\}$. You have to work with the vector equation $c_1\mathbf{v}_1 + \cdots + c_p\mathbf{v}_p = \mathbf{0}$, or with Theorem 5, unless the vectors happen to be n-tuples of numbers.

A set $\{\mathbf{v}\}$ with $\mathbf{v} \neq \mathbf{0}$ is linearly independent because the vector equation $x_1\mathbf{v} = \mathbf{0}$ has only the trivial solution. (See Exercise 28 in Section 5.1.) The set $\{\mathbf{0}\}$ is linearly dependent, because the equation $x_1\mathbf{0} = \mathbf{0}$ has many nontrivial solutions.

Theorem 7 is important for later work, but its proof is rather subtle. You will probably have to read page 215 several times to understand the principle involved.

A basis for a vector space V is a set in V that is large enough to span V and small enough to be linearly independent. See also "Two Views of a Basis" on page 216. The plural of *basis* is *bases*.

Warning: Many students confuse the idea of a set failing to span a space with the notion of linear dependence. If a set in V does not span V, the set may or may not be linearly dependent.

The table on the next page adds three new statements to the Invertible Matrix Theorem, at the bottom of the list from Section 3.3 in this Study Guide.

STATEMENTS FROM THE INVERTIBLE MATRIX THEOREM

Equivalent statements, for any $n \times p$ matrix A.	Equivalent only for an $n \times n$ square matrix A.	Equivalent statements, for any $n \times p$ matrix A.
f. There is a matrix C such that $AC = I$.	a. A is an invertible matrix.	g. There is a matrix D such that $DA = I$.
*. A has a pivot position in every row.	b′. A has n pivot positions.	*. A has a pivot position in every column.
e″. The columns of A span $\mathbb{R}^n$.	b. A is row equivalent to the $n \times n$ identity matrix.	d′. The columns of A form a linearly independent set.
*. The equation $A\mathbf{x} = \mathbf{b}$ has at least one solution for each $\mathbf{b}$ in $\mathbb{R}^n$.	e′. The equation $A\mathbf{x} = \mathbf{b}$ has a unique solution for each $\mathbf{b}$ in $\mathbb{R}^n$.	*. The equation $A\mathbf{x} = \mathbf{b}$ has at most one solution for each $\mathbf{b}$ in $\mathbb{R}^n$.
e‴. The linear transformation $\mathbf{x} \mapsto A\mathbf{x}$ maps onto $\mathbb{R}^n$.	*. The linear transformation $\mathbf{x} \mapsto A\mathbf{x}$ is invertible.	d″. The linear transformation $\mathbf{x} \mapsto A\mathbf{x}$ is one-to-one.
	c. A is a product of elementary matrices.	d. The equation $A\mathbf{x} = \mathbf{0}$ has only the trivial solution.
	h. A^T is an invertible matrix.	*. The equation $A\mathbf{x} = \mathbf{0}$ has no free variables.
*. Col $A = \mathbb{R}^n$	i. The columns of A form a basis of $\mathbb{R}^n$.	*. Nul $A = \{\mathbf{0}\}$

SOLUTIONS TO EXERCISES

1. The two vectors, call them $\mathbf{v}_1, \mathbf{v}_2$, are not multiples. (More precisely, neither vector is a multiple of the other.) Hence they are linearly independent. In fact, $\{\mathbf{v}_1, \mathbf{v}_2\}$ is a basis for $\mathbb{R}^2$, because the 2×2 matrix $A = [\mathbf{v}_1 \;\; \mathbf{v}_2]$ has linearly independent columns, and hence its columns span $\mathbb{R}^2$, by the Invertible Matrix Theorem.

7. The 3 vectors are in $\mathbb{R}^3$. Check the (square) matrix whose columns are these vectors:

$$A = \begin{bmatrix} 1 & 3 & -3 \\ 0 & 2 & -5 \\ -2 & -4 & 1 \end{bmatrix} \sim \begin{bmatrix} 1 & 3 & -3 \\ 0 & 2 & -5 \\ 0 & 0 & 0 \end{bmatrix}$$

The matrix A is not invertible, so its columns cannot form a basis for $\mathbb{R}^3$; the columns are not linearly independent and they do not span $\mathbb{R}^3$.

Study Tip: Theorem 2 in Section 2.2 will help you decide when a set of vectors spans $\mathbb{R}^n$.

13. Completely row reduce the augmented matrix for the equation $A\mathbf{x} = \mathbf{0}$:

$$[A \quad \mathbf{0}] = \begin{bmatrix} 1 & 0 & -3 & 2 & 0 \\ 0 & 1 & -5 & 4 & 0 \\ 3 & -2 & 1 & -2 & 0 \end{bmatrix} \sim \begin{bmatrix} 1 & 0 & -3 & 2 & 0 \\ 0 & 1 & -5 & 4 & 0 \\ 0 & 0 & 0 & 0 & 0 \end{bmatrix}$$

$$\begin{aligned} x_1 \quad - 3x_3 + 2x_4 &= 0 \\ x_2 - 5x_3 + 4x_4 &= 0 \\ x_3,\ x_4 \text{ are free} \end{aligned}$$

The general solution of $A\mathbf{x} = \mathbf{0}$ is $\begin{bmatrix} 3x_3-2x_4 \\ 5x_3-4x_4 \\ x_3 \\ x_4 \end{bmatrix} = x_3 \underset{\uparrow\ \mathbf{u}}{\begin{bmatrix} 3 \\ 5 \\ 1 \\ 0 \end{bmatrix}} + x_4 \underset{\uparrow\ \mathbf{v}}{\begin{bmatrix} -2 \\ -4 \\ 0 \\ 1 \end{bmatrix}}$. Since $\{\mathbf{u}, \mathbf{v}\}$ is automatically linearly independent (by the remarks that follow Example 4 in Section 5.2), $\{\mathbf{u}, \mathbf{v}\}$ is a basis for Nul A.

19. Use the method of Practice Problem 2:

$$A = [\mathbf{v}_1 \cdots \mathbf{v}_5] = \begin{bmatrix} 1 & 0 & -3 & 1 & 2 \\ 0 & 1 & -4 & -3 & 1 \\ -3 & 2 & 1 & -8 & -6 \\ 2 & -3 & 6 & 7 & 9 \end{bmatrix} \sim \begin{bmatrix} 1 & 0 & -3 & 1 & 2 \\ 0 & 1 & -4 & -3 & 1 \\ 0 & 2 & -8 & -5 & 0 \\ 0 & -3 & 12 & 5 & 5 \end{bmatrix}$$

$$\sim \begin{bmatrix} 1 & 0 & -3 & 1 & 2 \\ 0 & 1 & -4 & -3 & 1 \\ 0 & 0 & 0 & 1 & -2 \\ 0 & 0 & 0 & -4 & 8 \end{bmatrix} \sim \begin{bmatrix} 1 & 0 & -3 & 1 & 2 \\ 0 & 1 & -4 & -3 & 1 \\ 0 & 0 & 0 & 1 & -2 \\ 0 & 0 & 0 & 0 & 0 \end{bmatrix}$$

The pivot columns are 1,2, and 4. A basis for Col A = Span$\{\mathbf{v}_1, \ldots, \mathbf{v}_5\}$ is $\{\mathbf{v}_1, \mathbf{v}_2, \mathbf{v}_4\}$. Don't forget that you cannot, in general, use columns of the reduced matrix for the basis.

Warning: Exercises such as 17-20 are good exam questions. When some students encounter a question like Exercise 19, they are unsure whether they should use a null space or a column space technique. Such students have simply memorized *procedures* and they don't really know the *definitions* of Nul A and Col A. Learn the definitions *and* the procedures **now**. Otherwise, you are likely to have trouble with Section 5.6.

23. Let $A = [\mathbf{v}_1 \ \ \mathbf{v}_2 \ \ \mathbf{v}_3 \ \ \mathbf{v}_4]$. Since A is square and its columns span $\mathbb{R}^4$, the columns of A must be linearly independent, by the Invertible Matrix Theorem. So $\{\mathbf{v}_1, \mathbf{v}_2, \mathbf{v}_3, \mathbf{v}_4\}$ is a basis for $\mathbb{R}^4$.

25. The displayed equation shows only that Span$\{\mathbf{v}_1, \mathbf{v}_2, \mathbf{v}_3\}$ *contains* H. In fact, the vectors $\mathbf{v}_1, \mathbf{v}_2, \mathbf{v}_3$ are not all in H, so Span$\{\mathbf{v}_1, \mathbf{v}_2, \mathbf{v}_3\}$ cannot be H. Therefore $\{\mathbf{v}_1, \mathbf{v}_2, \mathbf{v}_3\}$ cannot be a basis for H. (It is easy to check that $\{\mathbf{v}_1, \mathbf{v}_2, \mathbf{v}_3\}$ is a basis for $\mathbb{R}^3$.)

31. (This generalizes Exercise 25 in Section 2.5.) Suppose $\{\mathbf{v}_1, \dots, \mathbf{v}_p\}$ is linearly dependent. Then there exist $c_1, \dots, c_p$, not all zero, such that

$$c_1\mathbf{v}_1 + \cdots + c_p\mathbf{v}_p = \mathbf{0}$$

Then, since T is linear,

$$T(c_1\mathbf{v}_1 + \cdots + c_p\mathbf{v}_p) = T(\mathbf{0}) = \mathbf{0}$$

and

$$c_1T(\mathbf{v}_1) + \cdots + c_pT(\mathbf{v}_p) = \mathbf{0}$$

Since not all the c_i are zero, $\{T(\mathbf{v}_1), \dots, T(\mathbf{v}_p)\}$ is linearly dependent.

Study Tip: The solution of Exercise 31 illustrates how to use linear dependence: If $\{\mathbf{v}_1, \dots, \mathbf{v}_p\}$ is known to be linearly dependent, then you can write $c_1\mathbf{v}_1 + \cdots + c_p\mathbf{v}_p = \mathbf{0}$, assume that not all the c_k are zero, and use this equation in some way.

MATLAB By now, the row reduction algorithm should be second nature, so it's time to use the power of MATLAB. The command **ref(A)** produces the reduced echelon form of A. From that you will immediately be able to write a basis for Col A and to write the homogeneous equations that describe Nul A. (Don't forget that A is a coefficient matrix, not an augmented matrix.)

If you are given vectors, say $\mathbf{v}_1$, $\mathbf{v}_2$, $\mathbf{v}_3$, apply **ref([v1 v2 v3])** to locate the pivot positions in the matrix $A = [\mathbf{v}_1 \ \ \mathbf{v}_2 \ \ \mathbf{v}_3]$. That information should tell you whether the vectors span $\mathbb{R}^n$ and whether the vectors are linearly independent.

I added the command **ref** to the Toolbox for the text in June, 1994, after I prepared the data disk for instructors. If you have the Toolbox disk, please share your copy with your instructor if needed. Note: MATLAB has another command, **rref** , that works basically the same as **ref** but seems to be slower. If you are interested in how **ref** works, you can examine the file `ref.m` with a text editor.

5.4 COORDINATE SYSTEMS

This section contains a variety of geometric and algebraic explanations of the idea of a coordinate system for a vector space.

KEY IDEAS

The coordinate mapping from a vector space V (with a basis of n elements) onto $\mathbb{R}^n$ is a rule for giving "$\mathbb{R}^n$-names" to vectors in V in such a way that the vector space structure of V is still visible in $\mathbb{R}^n$. Every vector space calculation in V is precisely mirrored by the same calculation in $\mathbb{R}^n$.

An important special case is when V is itself $\mathbb{R}^n$, and each vector $\mathbf{x}$ and its coordinate vector $[\mathbf{x}]_{\mathcal{B}}$ are related by a matrix equation $\mathbf{x} = P_{\mathcal{B}}[\mathbf{x}]_{\mathcal{B}}$.

Everything in the section depends on the Unique Representation Theorem. The proof of that theorem is important (and could appear on an exam) because it shows precisely why the two properties of a basis $\mathcal{B}$ are important, and it illustrates how linear independence can be used in an argument:

> Any vector in V has "coordinates" because $\mathcal{B}$ spans V, and the coordinates are uniquely determined because $\mathcal{B}$ is linearly independent.

If you are asked to prove the theorem, make sure your proof shows exactly where each property of a basis is needed in the proof. Also, be careful *not* to use a matrix in the proof. The vectors $\mathbf{v}_1, \ldots, \mathbf{v}_n$ cannot be arranged as the columns of an ordinary matrix when the vectors are in some abstract vector space.

Checkpoint: Let $\mathcal{B} = \{\mathbf{b}_1, \ldots, \mathbf{b}_n\}$ be a basis for $\mathbb{R}^n$. Apply the Invertible Matrix Theorem to the matrix $A = [\mathbf{b}_1 \ \cdots \ \mathbf{b}_n]$ and deduce the Unique Representation Theorem for the case when $V = \mathbb{R}^n$.

STUDY NOTES

Be careful to distinguish between $\mathbf{x}$ and $[\mathbf{x}]_{\mathcal{B}}$. They are *not equal* in general, even if $\mathbf{x}$ itself is in $\mathbb{R}^n$ (unless $\mathcal{B}$ is the standard basis for $\mathbb{R}^n$).

Theorem 9 and Exercises 25 and 26 show that the coordinate mapping translates vector space statements or calculations in V into equivalent (and familiar) calculations in $\mathbb{R}^n$. The table below lists some examples of typical linear algebra statements.

CORRESPONDING STATEMENTS IN ISOMORPHIC VECTOR SPACES

	Linear Algebra in V	Matrix Algebra in $\mathbb{R}^n$
a.	$\mathbf{u}$, $\mathbf{v}$, and $\mathbf{w}$ are in V	$[\mathbf{u}]_{\mathcal{B}}$, $[\mathbf{v}]_{\mathcal{B}}$, and $[\mathbf{w}]_{\mathcal{B}}$ are in $\mathbb{R}^n$
b.	$\mathbf{w}$ is in Span$\{\mathbf{u},\mathbf{v}\}$, or $\mathbf{w}$ is in the subspace of V spanned by $\mathbf{u}$ and $\mathbf{v}$	$[\mathbf{w}]_{\mathcal{B}}$ is in Span$\{[\mathbf{u}]_{\mathcal{B}}, [\mathbf{v}]_{\mathcal{B}}\}$, or $[\mathbf{w}]_{\mathcal{B}}$ is in the subspace of $\mathbb{R}^n$ spanned by $[\mathbf{u}]_{\mathcal{B}}$ and $[\mathbf{v}]_{\mathcal{B}}$
c.	$\mathbf{w} = c\mathbf{u} + d\mathbf{v}$	$[\mathbf{w}]_{\mathcal{B}} = c[\mathbf{u}]_{\mathcal{B}} + d[\mathbf{v}]_{\mathcal{B}}$
d.	$\{\mathbf{v}_1,\dots,\mathbf{v}_p\}$ is lin. indep.	$\{[\mathbf{v}_1]_{\mathcal{B}},\dots,[\mathbf{v}_p]_{\mathcal{B}}\}$ is lin. indep.
e.	$\{\mathbf{v}_1,\dots,\mathbf{v}_p\}$ spans V	$\{[\mathbf{v}_1]_{\mathcal{B}},\dots,[\mathbf{v}_p]_{\mathcal{B}}\}$ spans $\mathbb{R}^n$
f.	$\{\mathbf{v}_1,\dots,\mathbf{v}_n\}$ is a basis for V	$\{[\mathbf{v}_1]_{\mathcal{B}},\dots,[\mathbf{v}_n]_{\mathcal{B}}\}$ is a basis for $\mathbb{R}^n$

SOLUTIONS TO EXERCISES

1. Since $[\mathbf{x}]_{\mathcal{B}} = \begin{bmatrix}5\\3\end{bmatrix}$, we have $\mathbf{x} = 5\mathbf{b}_1 + 3\mathbf{b}_2 = 5\begin{bmatrix}3\\-5\end{bmatrix} + 3\begin{bmatrix}-4\\6\end{bmatrix} = \begin{bmatrix}3\\-7\end{bmatrix}$.

7. The $\mathcal{B}$-coordinates of $\mathbf{x}$ are scalars c_1, c_2, c_3 that satisfy $c_1\mathbf{b}_1 + c_2\mathbf{b}_2 + c_3\mathbf{b}_3 = \mathbf{x}$. To solve this vector equation, row reduce the augmented matrix:

$$\begin{bmatrix}1 & -3 & 2 & 8\\-1 & 4 & -2 & -9\\-3 & 9 & 4 & 6\end{bmatrix} \sim \begin{bmatrix}1 & -3 & 2 & 8\\0 & 1 & 0 & -1\\0 & 0 & 10 & 30\end{bmatrix} \sim \begin{bmatrix}1 & -3 & 0 & 2\\0 & 1 & 0 & -1\\0 & 0 & 1 & 3\end{bmatrix} \sim \begin{bmatrix}1 & 0 & 0 & -1\\0 & 1 & 0 & -1\\0 & 0 & 1 & 3\end{bmatrix}$$

$$\uparrow \quad \uparrow \quad \uparrow \quad \uparrow$$
$$\mathbf{b}_1 \quad \mathbf{b}_2 \quad \mathbf{b}_3 \quad \mathbf{x}$$

So $[\mathbf{x}]_{\mathcal{B}} = \begin{bmatrix}-1\\-1\\3\end{bmatrix}$.

13. Using the method of Practice Problem 2, we seek c_1, c_2, c_3 such that

$$c_1(1 + t^2) + c_2(t + t^2) + c_3(1 + 2t + t^2) = \mathbf{p}(t) = 1 + 4t + 7t^2 \quad (*)$$

Multiplying out terms on the left and equating coefficients of like powers of t, we obtain the system

$$\begin{aligned} c_1 \quad\quad + c_3 &= 1\\ c_2 + 2c_3 &= 4\\ c_1 + c_2 + c_3 &= 7 \end{aligned}$$

Row reduction of the augmented matrix produces

$$\begin{bmatrix} 1 & 0 & 1 & 1 \\ 0 & 1 & 2 & 4 \\ 1 & 1 & 1 & 7 \end{bmatrix} \sim \begin{bmatrix} 1 & 0 & 1 & 1 \\ 0 & 1 & 2 & 4 \\ 0 & 0 & -2 & 2 \end{bmatrix} \sim \begin{bmatrix} 1 & 0 & 0 & 2 \\ 0 & 1 & 0 & 6 \\ 0 & 0 & 1 & -1 \end{bmatrix}, \text{ and } [\mathbf{p}]_{\mathcal{B}} = \begin{bmatrix} 2 \\ 6 \\ -1 \end{bmatrix}$$

A slightly faster solution uses Theorem 9 and the fact that a calculation in $\mathbb{P}_2$ can be done instead with coordinate vectors of all the given polynomials relative to the standard basis $\{1, t, t^2\}$. Equation (*) in such coordinate vectors becomes

$$c_1\begin{bmatrix} 1 \\ 0 \\ 1 \end{bmatrix} + c_2\begin{bmatrix} 0 \\ 1 \\ 1 \end{bmatrix} + c_3\begin{bmatrix} 1 \\ 2 \\ 1 \end{bmatrix} = \begin{bmatrix} 1 \\ 4 \\ 7 \end{bmatrix}$$

This vector equation is, of course, equivalent to the system of equations above, and it is solved by the same row reduction process.

19. The set S spans V because every $\mathbf{x}$ in V has a representation as a (unique) linear combination of elements of S. To show linear independence, suppose that $S = \{\mathbf{v}_1, \ldots, \mathbf{v}_n\}$ and $c_1\mathbf{v}_1 + \cdots + c_n\mathbf{v}_n = \mathbf{0}$ for some scalars $c_1, \ldots, c_n$. The case when $c_1 = \cdots = c_n = 0$ is one possibility. By hypothesis, this is the *only* possible representation of the zero vector as a linear combination of the elements of S. So S is linearly independent and hence is a basis for V.

21. We want A to satisfy $A\mathbf{x} = [\mathbf{x}]_{\mathcal{B}}$ for each $\mathbf{x}$, and we know that the change-of-coordinates matrix satisfies $P_{\mathcal{B}}[\mathbf{x}]_{\mathcal{B}} = \mathbf{x}$ for each $\mathbf{x}$. Comparing these two equations, we see that

$$A = P_{\mathcal{B}}^{-1} = [\mathbf{b}_1 \;\; \mathbf{b}_2]^{-1} = \begin{bmatrix} 1 & -2 \\ -4 & 9 \end{bmatrix}^{-1} = \begin{bmatrix} 9 & 2 \\ 4 & 1 \end{bmatrix}$$

23. Suppose that $[\mathbf{u}]_{\mathcal{B}} = [\mathbf{w}]_{\mathcal{B}}$ for some $\mathbf{u}$ and $\mathbf{w}$ in V, and denote the entries in this coordinate vector by $c_1, \ldots, c_n$. By definition of the coordinate vectors,

$$\mathbf{u} = c_1\mathbf{b}_1 + \cdots + c_n\mathbf{b}_n \quad \text{and} \quad \mathbf{w} = c_1\mathbf{b}_1 + \cdots + c_n\mathbf{b}_n$$

which shows that $\mathbf{u} = \mathbf{w}$. Since $\mathbf{u}$ and $\mathbf{w}$ were arbitrary elements of V, this shows that the coordinate mapping is one-to-one.

25. Since the coordinate mapping is one-to-one, the following equations have the same solutions, $c_1, \dots, c_p$:

$$c_1\mathbf{u}_1 + \cdots + c_p\mathbf{u}_p = \mathbf{0} \qquad \text{(the zero vector in } V) \qquad (6)$$

$$[c_1\mathbf{u}_1 + \cdots + c_p\mathbf{u}_p]_{\mathcal{B}} = [\mathbf{0}]_{\mathcal{B}} \qquad \text{(the zero vector in } \mathbb{R}^n) \qquad (7)$$

Since the coordinate mapping is linear, (7) is equivalent to

$$c_1[\mathbf{u}_1]_{\mathcal{B}} + \cdots + c_p[\mathbf{u}_p]_{\mathcal{B}} = \begin{bmatrix} 0 \\ \vdots \\ 0 \end{bmatrix} \qquad (8)$$

Hence $c_1, \dots, c_p$ satisfy (6) if and only if they satisfy (8). So (6) has only the trivial solution if and only if (8) has only the trivial solution. It follows that $\{\mathbf{u}_1, \dots, \mathbf{u}_p\}$ is linearly independent if and only if $\{[\mathbf{u}_1]_{\mathcal{B}}, \dots, [\mathbf{u}_p]_{\mathcal{B}}\}$ is linearly independent. (This fact is also an immediate consequence of Exercises 31 and 32 in Section 5.3.)

Warning: The standard mistake in Exercises 27-30 is to write the coordinate vectors as the *rows* of a matrix and then to turn around and check the linear independence or dependence of the *columns*. Since we mainly work with column vectors, it is wise to write the coordinate vectors first (as columns) and afterwards write a matrix that can be row reduced to check the linear independence of its columns.

27. The coordinate vectors of the polynomials are $\begin{bmatrix} 1 \\ 2 \\ 0 \end{bmatrix}, \begin{bmatrix} 3 \\ -1 \\ 0 \end{bmatrix}, \begin{bmatrix} -1 \\ 0 \\ 3 \end{bmatrix}$ (relative to the standard basis). These vectors are easily seen to be linearly independent. (For instance, you can use Theorem 5. In fact, you could even use Theorem 5 easily on the polynomials themselves.)

29. Use the same method as in Exercise 27. Show that $\begin{bmatrix} 1 \\ 0 \\ 0 \\ 1 \end{bmatrix}, \begin{bmatrix} 3 \\ 1 \\ -2 \\ 0 \end{bmatrix}, \begin{bmatrix} 0 \\ -1 \\ 3 \\ -1 \end{bmatrix}$ are linearly independent in $\mathbb{R}^4$. This will show that the corresponding polynomials in $\mathbb{P}^3$ are linearly independent.

Study Tip: Exercises 27—30 are easily changed into a question whether the given polynomials form a basis for $\mathbb{P}_3$. What else would you have to write or calculate? Remember, that although you may look at vectors in $\mathbb{R}^4$, your answer must include a discussion about the polynomials themselves. That is, you must explain why the given polynomials form, or do not form, a basis for the space of polynomials.

Answer to Checkpoint: The columns of the matrix $A = [\mathbf{b}_1 \ \cdots \ \mathbf{b}_n]$ form a basis for $\mathbb{R}^n$. By statements (i) and (e′) of the Invertible Matrix Theorem (in the previous section), for each $\mathbf{x}$ in $\mathbb{R}^n$ there exists a unique vector $\mathbf{c} = (c_1, \ldots, c_n)$ such that $\mathbf{x} = A\mathbf{c} = c_1\mathbf{b}_1 + \cdots + c_n\mathbf{b}_n$.

MATLAB If an equation $A\mathbf{x} = \mathbf{b}$ has a unique solution, MATLAB will automatically produce $\mathbf{x}$ if you use the command

x = A\b

In this section, the equation probably will have the form $P\mathbf{u} = \mathbf{x}$, with $\mathbf{u}$ the $\mathcal{B}$-coordinate vector of $\mathbf{x}$, and the command will be **u = P\x**.

The "backslash" command works in two different ways. When A is square, the command **A\b** causes MATLAB to create an LU factorization of A (see Section 3.5); if A is invertible, the factorization is used to produce the unique solution to $A\mathbf{x} = \mathbf{b}$; and if A is not invertible, MATLAB gives the error message "*matrix is singular*" (even if the system $A\mathbf{x} = \mathbf{b}$ has a solution). When A is not square, **A\b** creates a least-squares solution (see Section 7.5).

5.5 THE DIMENSION OF A VECTOR SPACE

This short section provides a convenient way to compare the "sizes" of various subspaces of a vector space without having first to choose specific bases for the subspaces and then to count the number of basis elements.

KEY IDEAS

Theorem 11 shows that the dimension of a finite-dimensional vector space is well-defined and does not depend on the particular basis for the space.

Theorems 10 and 13 are important for later theory and applications. You might remember Theorem 10 more easily in this form:

> In an n-dimensional vector space, any set of more than n vectors must be linearly dependent.

You already know this, of course, for $\mathbb{R}^n$. And if you read about isomorphisms in Section 5.4, then you can see why the result is true in general, because any n-dimensional vector space is isomorphic to $\mathbb{R}^n$.

Theorem 13 may be restated as follows:

> If $\dim V = n \geq 1$ and if S is a subset of V that contains precisely n elements, then S is linearly independent if and only if S spans V.

Warning: Suppose H is a subspace of a finite-dimensional space V. Theorem 12 implies that any basis of H can be extended to a basis of V. But it is *not* true that any basis of V can be cut down to a basis for H. That is, if S is a basis for V, it is not likely that a subset of S is a basis for H. For instance, consider the standard basis for $\mathbb{R}^3$ and any plane H that contains the origin but none of the coordinate axes.

SOLUTIONS TO EXERCISES

1. Since $\begin{bmatrix} s - 2t \\ s + t \\ 3t \end{bmatrix} = s\begin{bmatrix} 1 \\ 1 \\ 0 \end{bmatrix} + t\begin{bmatrix} -2 \\ 1 \\ 3 \end{bmatrix}$ for all s, t, the set $\left\{\begin{bmatrix} 1 \\ 1 \\ 0 \end{bmatrix}, \begin{bmatrix} -2 \\ 1 \\ 3 \end{bmatrix}\right\}$ certainly spans the subspace, call it H. Also, the set is obviously linearly independent (because the vectors are not multiples), so the set is a basis for H. Hence, $\dim H = 2$.

3. The given subspace, call it H, is the set of all linear combinations of the vectors

$$\mathbf{v}_1 = \begin{bmatrix} 0 \\ 1 \\ 0 \\ 1 \end{bmatrix}, \quad \mathbf{v}_2 = \begin{bmatrix} 0 \\ -1 \\ 1 \\ 2 \end{bmatrix}, \quad \mathbf{v}_3 = \begin{bmatrix} 2 \\ 0 \\ -3 \\ 0 \end{bmatrix}$$

First determine if $\{\mathbf{v}_1, \mathbf{v}_2, \mathbf{v}_3\}$ is linearly independent. One way to do this is to row reduce the augmented matrix $[\mathbf{v}_1 \ \ \mathbf{v}_2 \ \ \mathbf{v}_3 \ \ \mathbf{0}]$. A faster way is to use Theorem 5 in Section 5.3. Clearly, $\mathbf{v}_2$ is not a multiple of $\mathbf{v}_1$, and $\mathbf{v}_3$ is not a linear combination of the vectors $\mathbf{v}_1, \mathbf{v}_2$ that precede it, because the first entry in $\mathbf{v}_3$ is not zero. Hence $\{\mathbf{v}_1, \mathbf{v}_2, \mathbf{v}_3\}$ is linearly independent and thus is a basis for the space H it spans. Thus $\dim H = 3$.

7. The set of solutions of the homogeneous system consists of only the trivial solution. So the subspace is $\{\mathbf{0}\}$, and it has no basis. (The vector $\mathbf{0}$ spans the space, but $\{\mathbf{0}\}$ is a linearly dependent set.) By definition, the dimension is zero. [*Note*: People who want every subspace to have a basis often define the empty set to be a basis for $\{\mathbf{0}\}$. The number of vectors in this basis is zero, so the dimension of $\{\mathbf{0}\}$ is still zero.]

13. A has three pivot columns, so $\dim \text{Col } A = 3$. There are two columns without pivot positions, so the equation $A\mathbf{x} = \mathbf{0}$ has two free variables, and $\dim \text{Nul } A = 2$.

19. Form the matrix whose columns are the coordinate vectors of the Hermite polynomials, relative to the standard basis $\{1, t, t^2, t^3\}$:

$$A = \begin{bmatrix} 1 & 0 & -2 & 0 \\ 0 & 2 & 0 & -12 \\ 0 & 0 & 4 & 0 \\ 0 & 0 & 0 & 8 \end{bmatrix}$$

The matrix has four pivots and hence is invertible. So its columns, the coordinate vectors, are linearly independent. Hence the Hermite polynomials themselves are linearly independent in $\mathbb{P}_3$. Since there are *four* Hermite polynomials, and dim $\mathbb{P}_3 = 4$, we conclude from Theorem 13 that the Hermite polynomials form a *basis* for $\mathbb{P}_3$.

Note: You could, of course, say that the columns of the matrix A span $\mathbb{R}^4$. But you cannot stop with that assertion, because you need the polynomials to span $\mathbb{P}_3$. You have to go on and point out that because of the isomorphism between $\mathbb{P}_3$ and $\mathbb{R}^4$, a set of vectors spans $\mathbb{P}_3$ if and only if the set of coordinate vectors (the columns of A) spans $\mathbb{R}^4$. So the solution is shorter if you appeal to Theorem 13.

23. Note that $n \geq 1$, because S cannot have fewer than 0 vectors. If dim $V = n \geq 1$, then $V \neq \{\mathbf{0}\}$. If S spans V, then a subset S' of S is a basis for V, by the Spanning Set Theorem. But if S has fewer than n vectors, then S' also has fewer than n vectors. This is impossible, by Theorem 11, because dim $V = n$. So S cannot span V.

25. If $\mathbb{P}$ were finite-dimensional, then Theorem 12 would imply that $n + 1 = \dim \mathbb{P}_n \leq \dim \mathbb{P}$ for each n, because each $\mathbb{P}_n$ is a subspace of $\mathbb{P}$. This is impossible, so $\mathbb{P}$ must be infinite-dimensional.

27. True. Apply the Spanning Set Theorem to the set $\{\mathbf{v}_1, \ldots, \mathbf{v}_p\}$ and produce a basis for V. This basis will have no more than p elements in it, so dim V must be no more than p. (In the special case when $\mathbf{v}_1 = \cdots = \mathbf{v}_p = \mathbf{0}$, the space V is itself the zero vector space and its dimension does not exceed p.)

29. True. By Theorem 12, $\{\mathbf{v}_1, \ldots, \mathbf{v}_p\}$ can be expanded to a basis for V. The basis will have at least p elements in it, so dim V must be at least p.

31. True. Take any basis (of p vectors) for V and adjoin the zero vector. Spanning sets can be arbitrarily large. The dimension of V being p only keeps spanning sets from having *fewer* than p elements.

33. Since H is a subspace of a finite-dimensional space, H is finite-dimensional. If $H = \{\mathbf{0}\}$, then $T(H)$ is the zero vector and dim $T(H)$ = dim H. Otherwise, H has a basis, say, $\mathbf{v}_1,\dots,\mathbf{v}_p$. Any vector in $T(H)$ has the form $T(\mathbf{y})$ for some $\mathbf{y}$ in H. Since $\{\mathbf{v}_1,\dots,\mathbf{v}_p\}$ spans H, there exist scalars $c_1,\dots,c_p$ such that $\mathbf{y} = c_1\mathbf{v}_1 + \cdots + c_p\mathbf{v}_p$. Since T is linear, $T(\mathbf{y}) = c_1T(\mathbf{v}_1) + \cdots + c_pT(\mathbf{v}_p)$. This shows that $\{T(\mathbf{v}_1),\dots, T(\mathbf{v}_p)\}$ spans $T(H)$. By Exercise 27, dim $T(H) \le p$ = dim H.

Hint for Exercise 34: Use an exercise in Section 5.3.

Warning: The next section is quite important. Do your best to get caught up now. Otherwise, you may have difficulty relating the various concepts and facts about matrices that will be reviewed in Section 5.6.

Appendix: Bases for Subspaces of $\mathbb{R}^p$

If you have skipped most or all of Section 5.4 and are focusing on subspaces of $\mathbb{R}^n$ rather than general vector spaces, then the following result can take the place of Theorem 11.

Theorem. Any two bases of a subspace V of $\mathbb{R}^p$ contain the same number of elements.

Proof. Suppose that one basis of V is $\{\mathbf{b}_1,\dots,\mathbf{b}_m\}$ and another is $\{\mathbf{c}_1,\dots,\mathbf{c}_n\}$, with $m < n$. We shall show that this leads to a contradiction. Since the columns of the matrix $B = [\mathbf{b}_1 \ \cdots \ \mathbf{b}_m]$ form a basis of V, these columns span V. Hence for each $\mathbf{c}_j$ there is a vector $\mathbf{a}_j$ of weights such that

$$B\mathbf{a}_j = \mathbf{c}_j \qquad (j = 1,\dots,n) \qquad (*)$$

Note that each $\mathbf{a}_j$ is in $\mathbb{R}^m$, because B has m columns. Let $A = [\mathbf{a}_1 \ \cdots \ \mathbf{a}_n]$ and $C = [\mathbf{c}_1 \ \cdots \ \mathbf{c}_n]$. The equations in (*) show that $BA = C$. Since A is $m \times n$, with more columns than rows, there exists a nonzero $\mathbf{x}$ such that $A\mathbf{x} = \mathbf{0}$. Hence $C\mathbf{x} = BA\mathbf{x} = B\mathbf{0} = \mathbf{0}$. This shows that the columns of C are linearly dependent, which contradicts the fact that the columns of C form a basis for V. We conclude that m and n cannot be unequal. So the two bases have the same number of elements.

SECTION 5.6 RANK

This section gives you a chance to put together most of the ideas of the chapter in the same way that Section 3.3 collected the main ideas of the first three chapters.

KEY IDEAS

The Rank Theorem is the main result. By definition, rank A = dim Col A. But because rank A is also the dimension of Row A, the displayed equation in the theorem leads to the equation: dim Row A + dim Nul $A = n$.

> **Equivalent Descriptions of Rank**
>
> The rank of an $m \times n$ matrix A may be described in several ways:
>
> ▸the dimension of the column space of A, (our definition)
> ▸the number of pivot columns in A, (from Theorem 7)
> ▸the maximum number of linearly independent columns in A,
> ▸the dimension of the row space of A, (from the Rank Theorem)
> ▸the maximum number of linearly independent rows in A,
> ▸the number of nonzero rows in an echelon form of A,
> ▸the maximum number of columns in an invertible submatrix of A.
> (Supplementary Exercise 12 at the end of the chapter)

Pay attention to how Theorem 14 differs from the results in Section 5.3 about Col A: If you are interested in *rows* of A, use the nonzero rows of an echelon form B as a basis for Row A; if you are interested in the *columns* of A, only use B to obtain *information* about A (namely, to identify the pivot columns), and use the pivot columns of A as a basis for Col A. For Nul A, it is important to use the *reduced* echelon form of A.

When a matrix A is changed into a matrix B by one or more elementary row operations, the row space, null space, and column space of A may or may not be the same as the corresponding subspaces for B. The following table summarizes what can happen in this situation.

> **Effects of Elementary Row Operations**
>
> ▸Row operations do not affect the linear dependence relations among the columns. (That is, the columns of A have exactly the same linear dependence relations as the columns of any matrix row-equivalent to A.)
> ▸Row operations usually change the column space.
> ▸Row operations never change the row space.
> ▸Row operations never change the null space.

The four subspaces shown in Figure 1 in the text are called the *fundamental subspaces* associated with A. (See Exercises 27-29.) The main difficulty here is to avoid confusion between Row A, Nul A, and Col A. The fourth subspace will appear again in Sections 7.1 and 8.4.

The table below adds three more statements to the bottom of the list for the Invertible Matrix Theorem, Section 5.3 in this Study Guide.

STATEMENTS FROM THE INVERTIBLE MATRIX THEOREM

Equivalent statements, for any $n \times p$ matrix A.	Equivalent only for an $n \times n$ square matrix A.	Equivalent statements, for any $n \times p$ matrix A.
f. There is a matrix C such that $AC = I$.	a. A is an invertible matrix.	g. There is a matrix D such that $DA = I$.
*. A has a pivot position in every row.	b′. A has n pivot positions.	*. A has a pivot position in every column.
e″. The columns of A span $\mathbb{R}^n$.	b. A is row equivalent to the $n \times n$ identity matrix.	d′. The columns of A form a linearly independent set.
*. The equation $A\mathbf{x} = \mathbf{b}$ has at least one solution for each $\mathbf{b}$ in $\mathbb{R}^n$.	e′. The equation $A\mathbf{x} = \mathbf{b}$ has a unique solution for each $\mathbf{b}$ in $\mathbb{R}^n$.	*. The equation $A\mathbf{x} = \mathbf{b}$ has at most one solution for each $\mathbf{b}$ in $\mathbb{R}^n$.
e‴. The linear transformation $\mathbf{x} \mapsto A\mathbf{x}$ maps $\mathbb{R}^p$ onto $\mathbb{R}^n$.	*. The linear transformation $\mathbf{x} \mapsto A\mathbf{x}$ is invertible.	d″. The linear transformation $\mathbf{x} \mapsto A\mathbf{x}$ is one-to-one.
	c. A is a product of elementary matrices.	d. The equation $A\mathbf{x} = \mathbf{0}$ has only the trivial solution.
	h. A^T is an invertible matrix.	*. The equation $A\mathbf{x} = \mathbf{0}$ has no free variables.
j. Col $A = \mathbb{R}^n$	i. The columns of A form a basis of $\mathbb{R}^n$.	m. Nul $A = \{\mathbf{0}\}$
k. dim Col $A = n$	l. rank $A = n$.	n. dim Nul $A = 0$

SOLUTIONS TO EXERCISES

1. In the first printing of the text, the (2,4)-entry of the matrix B was 6 instead of -6, and the answers there were based on the incorrect B.

The solution below uses the corrected B.

$$A = \begin{bmatrix} 1 & -4 & 9 & -7 \\ -1 & 2 & -4 & 1 \\ 5 & -6 & 10 & 7 \end{bmatrix} \sim B = \begin{bmatrix} 1 & 0 & -1 & 5 \\ 0 & -2 & 5 & -6 \\ 0 & 0 & 0 & 0 \end{bmatrix}$$

Look at B, and conclude that A has two pivot columns and the equation $A\mathbf{x} = \mathbf{0}$ has two free variables. So rank $A = 2$ and dim Nul $A = 2$. In fact, the first two columns of A are pivot columns, so

$$\text{Basis for Col } A: \left\{ \begin{bmatrix} 1 \\ -1 \\ 5 \end{bmatrix}, \begin{bmatrix} -4 \\ 2 \\ -6 \end{bmatrix} \right\}$$

For the row space, use the rows in the echelon form B. That is,

Basis for Row A: $\{(1,0,-1,5),\ (0,-2,5,-6)\}$

For the null space, we need the reduced echelon form of A, and use it to solve $A\mathbf{x} = \mathbf{0}$:

$$A \sim B \sim \begin{bmatrix} 1 & 0 & -1 & 5 \\ 0 & 1 & -5/2 & 3 \\ 0 & 0 & 0 & 0 \end{bmatrix}; \qquad \begin{aligned} x_1 \qquad - \qquad x_3 + 5x_4 &= 0 \\ x_2 - (5/2)x_3 + 3x_4 &= 0 \\ 0 &= 0 \end{aligned}$$

Thus $x_1 = x_3 - 5x_4$, $x_2 = (5/2)x_3 - 3x_4$, with x_3, x_4 free. The general solution of $A\mathbf{x} = \mathbf{0}$ is

$$\begin{bmatrix} x_1 \\ x_2 \\ x_3 \\ x_4 \end{bmatrix} = \begin{bmatrix} x_3 - 5x_4 \\ (5/2)x_3 - 3x_4 \\ x_3 \\ x_4 \end{bmatrix} = x_3 \begin{bmatrix} 1 \\ 5/2 \\ 1 \\ 0 \end{bmatrix} + x_4 \begin{bmatrix} -5 \\ -3 \\ 0 \\ 1 \end{bmatrix}.$$

$$\uparrow \mathbf{u} \qquad\qquad \uparrow \mathbf{v}$$

Thus $\{\mathbf{u}, \mathbf{v}\}$ is a basis for Nul A.

Study Tip: Because rank $A = 2$ in Exercise 1, *any* two linearly independent columns of A form a basis for Col A, and any two linearly independent rows of A form a basis for Row A. When the rank of a matrix exceeds 2, selecting bases in this way is not so easy. (That is why you examine an echelon form of A.) On an exam, you should always choose the pivot columns of A as the basis for Col A and the nonzero rows of an echelon form of A as the basis for Row A. This will show that you can handle matrices with any rank.

7. Yes, Col $A = \mathbb{R}^4$, because Col A is a 4-dimensional subspace of $\mathbb{R}^4$ and hence coincides with $\mathbb{R}^4$. No, Nul A cannot be $\mathbb{R}^3$, because the vectors in Nul A have 7 entries. Nul A is a 3-dimensional subspace of $\mathbb{R}^7$, by the Rank Theorem.

13. The maximum rank of a 7×5 matrix A is 5 because A has only 5 columns to span Col A. The maximum rank of a 5×7 matrix is also 5, because even though A has 7 columns, the columns belong to $\mathbb{R}^5$ and hence can span at most a five-dimensional space.

19. The system may be represented as $A\mathbf{x} = \mathbf{0}$ where A is a 5×6 matrix. The solution space is one-dimensional. By the Rank Theorem, rank $A = 6 - 1 = 5$. So dim Col $A = 5$. But Col A is a subspace of $\mathbb{R}^5$. Hence Col $A = \mathbb{R}^5$. Thus $A\mathbf{x} = \mathbf{b}$ has a solution for all $\mathbf{b}$.

Study Tip: Exercises 19—25 make good exam questions.

23. The set of interest is the null space of a 12×8 matrix A. The description of this set suggests that dim Nul $A = 2$. By the Rank Theorem, rank $A = 8 - 2 = 6$. So the equation $A\mathbf{x} = \mathbf{0}$ is equivalent to $B\mathbf{x} = \mathbf{0}$, where B is an echelon form of A with 6 nonzero rows. The answer to the question is that six homogeneous equations are sufficient.

25. Let A be the 10×12 coefficient matrix. The general solution of the homogeneous system has the form $\mathbf{p} + c_1\mathbf{u}_1 + c_2\mathbf{u}_2 + c_3\mathbf{u}_3$. Assuming that the solution has been constructed in the standard way, $\mathbf{u}_1$, $\mathbf{u}_2$, $\mathbf{u}_3$ are linearly independent and span Nul A. (Said another way, there are three free variables in the system $A\mathbf{x} = \mathbf{b}$.) So dim Nul $A = 3$. By the Rank Theorem, dim Col $A = 12 - 3 = 9$. Since Col A is a subspace of $\mathbb{R}^{10}$ (because A has 10 rows), Col A cannot be all of $\mathbb{R}^{10}$, so some nonhomogeneous equations $A\mathbf{x} = \mathbf{b}$ will *not* have solutions.

31. The solution is in the text.

33. Let $A = [\mathbf{u} \quad \mathbf{u}_2 \quad \mathbf{u}_3]$. If $\mathbf{u} \neq \mathbf{0}$, then $\mathbf{u}$ must be a basis for Col A, since Col A is one-dimensional. Hence there exist scalars r and s such that $\mathbf{u}_2 = r\mathbf{u}$ and $\mathbf{u}_3 = s\mathbf{u}$, so that $A = [\mathbf{u} \quad r\mathbf{u} \quad s\mathbf{u}]$. If the entries in $\mathbf{u}$ are a, b, c, then

$$A = \begin{bmatrix} a & ra & sa \\ b & rb & sb \\ c & rc & sc \end{bmatrix} = \begin{bmatrix} a \\ b \\ c \end{bmatrix} [1 \quad r \quad s] = \mathbf{u}\mathbf{v}^T, \text{ where } \mathbf{v} = \begin{bmatrix} 1 \\ r \\ s \end{bmatrix}$$

If the first column of A is zero and the second column, call it $\mathbf{u}$, is nonzero, then $A = [\mathbf{0} \quad \mathbf{u} \quad r\mathbf{u}]$ for some r. In this case, take $\mathbf{v} = (0, 1, r)$. If $A = [\mathbf{0} \quad \mathbf{0} \quad \mathbf{u}]$, take $\mathbf{v} = (0, 0, 1)$.

MATLAB The command **rank(A)** is based on the singular value decomposition (Section 8.4). See page 5-10 for the command **ref(A)** .

5.7 CHANGE OF BASIS

This section will help you better understand coordinate systems. A review of Section 5.4 now is strongly recommended.

KEY IDEAS

Figure 1 and the accompanying discussion will help you visualize the main idea of the section. Imagine superimposing the $\mathcal{C}$-graph paper (Figure 1-b) on the $\mathcal{B}$-graph paper (Figure 1-a). Can you see where $\mathbf{b}_1$ will lie on the $\mathcal{C}$-coordinate system? Four units in the $\mathbf{c}_1$-direction and one unit in the $\mathbf{c}_2$-direction. That is the geometric interpretation of the equation $[\mathbf{b}_1]_{\mathcal{C}} = \begin{bmatrix} 4 \\ 1 \end{bmatrix}$ in Example 1. Similarly, since $[\mathbf{b}_2]_{\mathcal{C}} = \begin{bmatrix} -6 \\ 1 \end{bmatrix}$, $\mathbf{b}_2$ lies six units in the negative $\mathbf{c}_1$-direction and one unit in the $\mathbf{c}_2$-direction.

In general, the locations of $\mathbf{b}_1$ and $\mathbf{b}_2$ on the $\mathcal{C}$-graph paper are precisely what you must find in order to build the columns of the change-of-coordinates matrix:

$$\underset{\mathcal{C}\leftarrow\mathcal{B}}{P} = [[\mathbf{b}_1]_{\mathcal{C}} \quad [\mathbf{b}_2]_{\mathcal{C}}]$$

The notation for this matrix should help you remember the basic equation for changing $\mathcal{B}$-coordinates into $\mathcal{C}$-coordinates:

$$[\mathbf{x}]_{\mathcal{C}} = \underset{\mathcal{C}\leftarrow\mathcal{B}}{P}[\mathbf{x}]_{\mathcal{B}}$$

The calculations are simple when $\mathcal{B}$ and $\mathcal{C}$ are bases for $\mathbb{R}^n$. The box after Example 2 illustrates the algorithm for computing the change-of-coordinates matrix. In general,

$$[\mathbf{c}_1 \ \cdots \ \mathbf{c}_n \ \vdots \ \mathbf{b}_1 \ \cdots \ \mathbf{b}_n] \sim [\ I \ \vdots \ \underset{\mathcal{C}\leftarrow\mathcal{B}}{P}\ ]$$

Equivalently, using the notation of Section 5.4,

$$[P_{\mathcal{C}} \quad P_{\mathcal{B}}] \sim [\ I \quad \underset{\mathcal{C}\leftarrow\mathcal{B}}{P}]$$

where $P_{\mathcal{B}}$ is the matrix $[\mathbf{b}_1 \ \cdots \ \mathbf{b}_n]$ that changes $\mathcal{B}$-coordinates to *standard coordinates*, and $P_{\mathcal{C}}$ is similarly defined. If you refer back to Exercise 9 of Section 3.2, you will see that $\underset{\mathcal{C}\leftarrow\mathcal{B}}{P}$ is the same as $(P_{\mathcal{C}})^{-1}P_{\mathcal{B}}$. Since $(P_{\mathcal{C}})^{-1}$ changes standard coordinates to $\mathcal{C}$-coordinates, we can obtain $[\mathbf{x}]_{\mathcal{C}}$ from $[\mathbf{x}]_{\mathcal{B}}$ as follows:

$$[\mathbf{x}]_{\mathcal{B}} \longrightarrow P_{\mathcal{B}}[\mathbf{x}]_{\mathcal{B}} \longrightarrow (P_{\mathcal{C}})^{-1}P_{\mathcal{B}}[\mathbf{x}]_{\mathcal{B}} = \underset{\mathcal{C}\leftarrow\mathcal{B}}{P}[\mathbf{x}]_{\mathcal{B}} = [\mathbf{x}]_{\mathcal{C}}$$

B-coordinates — standard coordinates — C-coordinates

This diagram provides another way of viewing the change of coordinates.

SOLUTIONS TO EXERCISES

1. a. From $\mathbf{b}_1 = 6\mathbf{c}_1 - 2\mathbf{c}_2$ and $\mathbf{b}_2 = 9\mathbf{c}_1 - 4\mathbf{c}_2$, write

$$[\mathbf{b}_1]_{\mathcal{C}} = \begin{bmatrix} 6 \\ -2 \end{bmatrix}, \quad \mathbf{b}_2 = \begin{bmatrix} 9 \\ -4 \end{bmatrix}, \text{ and } \underset{\mathcal{C}\leftarrow\mathcal{B}}{P} = \begin{bmatrix} 6 & 9 \\ -2 & -4 \end{bmatrix}$$

b. Since $\mathbf{x} = -3\mathbf{b}_1 + 2\mathbf{b}_2$,

$$[\mathbf{x}]_{\mathcal{B}} = \begin{bmatrix} -3 \\ 2 \end{bmatrix} \text{ and } [\mathbf{x}]_{\mathcal{C}} = \begin{bmatrix} 6 & 9 \\ -2 & -4 \end{bmatrix}\begin{bmatrix} -3 \\ 2 \end{bmatrix} = \begin{bmatrix} 0 \\ -2 \end{bmatrix}$$

7. Unlike Exercise 1, you do not have direct information from which you can write $[\mathbf{b}_1]_{\mathcal{C}}$ and $[\mathbf{b}_2]_{\mathcal{C}}$. Rather than compute these two coordinate vectors separately, use the algorithm from Example 2:

$$[\mathbf{c}_1 \ \ \mathbf{c}_2 \ \ \mathbf{b}_1 \ \ \mathbf{b}_2] = \begin{bmatrix} 1 & -2 & 7 & -3 \\ -5 & 2 & 5 & -1 \end{bmatrix} \sim \begin{bmatrix} 1 & -2 & 7 & -3 \\ 0 & -8 & 40 & -16 \end{bmatrix}$$

$$\sim \begin{bmatrix} 1 & -2 & 7 & -3 \\ 0 & 1 & -5 & 2 \end{bmatrix} \sim \begin{bmatrix} 1 & 0 & -3 & 1 \\ 0 & 1 & -5 & 2 \end{bmatrix}$$

Thus $\underset{\mathcal{C}\leftarrow\mathcal{B}}{P} = \begin{bmatrix} -3 & 1 \\ -5 & 2 \end{bmatrix}$. The change-of-coordinates matrix from $\mathcal{C}$ to $\mathcal{B}$ is

$$\underset{\mathcal{B}\leftarrow\mathcal{C}}{P} = (\underset{\mathcal{C}\leftarrow\mathcal{B}}{P})^{-1} = \begin{bmatrix} -3 & 1 \\ -5 & 2 \end{bmatrix}^{-1} = \frac{1}{-1}\begin{bmatrix} 2 & -1 \\ 5 & -3 \end{bmatrix} = \begin{bmatrix} -2 & 1 \\ -5 & 3 \end{bmatrix}$$

13. The complete answer is in the text.

MATLAB Data are available for Exercises 7-10. The command **ref(M)** will completely row reduce a matrix such as $M = [\mathbf{c}_1 \ \ \mathbf{c}_2 \ \ \mathbf{b}_1 \ \ \mathbf{b}_2]$ to the desired form.

5.8 APPLICATIONS TO DIFFERENCE EQUATIONS

Difference equations are the discrete analogues of differential equations. Both are important in science and engineering. The discrete and continuous theories are remarkably parallel, and linear algebra is applied in similar ways, although the calculations are somewhat easier for difference equations. A variety of examples and exercises here illustrate some difference equations you may encounter later in your work.

KEY IDEAS

Each signal in $\mathbb{S}$ is an infinite list of numbers. Linear independence of a set of signals can often be demonstrated by looking at short segments of the signals, that is, by showing that a Casorati matrix is invertible. A Casorati matrix cannot be used in general to demonstrate linear *dependence* of a set. However, see the appendix at the end of the section.

The main focus of the section is on difference equations. Given a homogeneous difference equation, you should be able to:

- ▸ determine whether a specified signal is a solution of the equation;
- ▸ find solutions of the equation, using the auxiliary equation;
- ▸ give the general solution (when the auxiliary equation has no multiple roots and no complex roots).

Finding solutions of a difference equation is not difficult. Showing that you have found *all* possible solutions would be extremely difficult were it not for Theorem 18. That is the main result of the section. Example 5 shows how to combine this theorem with Theorem 13 in Section 5.5.

The subsection on nonhomogenous equations is optional. If this is covered, you should be able to work Exercises 25-28. The general principle is illustrated in Figure 4, page 255:

$$\left\{\begin{matrix}\text{General solution of}\\ \text{nonhomogeneous eqn.}\end{matrix}\right\} = \left\{\begin{matrix}\text{Particular solution of}\\ \text{nonhomogeneous eqn.}\end{matrix}\right\} + \left\{\begin{matrix}\text{General solution of}\\ \text{homogeneous eqn.}\end{matrix}\right\}$$

The final subsection shows the modern way to study an *nth* order linear difference equation, rewriting it as a first order system $\mathbf{x}_{k+1} = A\mathbf{x}_k$ $(k = 1,2,\ldots)$. Such systems were introduced in Section 2.7 and they will be discussed further in Sections 5.9 and 6.6.

SOLUTIONS TO EXERCISES

1. If $y_k = (-4)^k$, then $y_{k+1} = (-4)^{k+1}$ and $y_{k+2} = (-4)^{k+2}$. Substitute these formulas into the left side of the equation:

$$\begin{aligned} y_{k+2} + 2y_{k+1} - 8y_k &= (-4)^{k+2} + 2(-4)^{k+1} - 8(-4)^k \\ &= (-4)^k[(-4)^2 + 2(-4) - 8] \\ &= (-4)^k[16 - 8 - 8] = 0 \quad \text{for all } k \end{aligned}$$

Since the difference equation holds for all k, $(-4)^k$ is a solution. The text answer displays the similar calculations for $y_k = 2^k$.

7. Compute the Casorati matrix for the signals 1^k, 2^k, and $(-2)^k$, setting $k = 0$ for convenience:

$$\begin{bmatrix} 1^0 & 2^0 & (-2)^0 \\ 1^1 & 2^1 & (-2)^1 \\ 1^2 & 2^2 & (-2)^2 \end{bmatrix} = \begin{bmatrix} 1 & 1 & 1 \\ 1 & 2 & -2 \\ 1 & 4 & 4 \end{bmatrix} \sim \begin{bmatrix} 1 & 1 & 1 \\ 0 & 1 & -3 \\ 0 & 3 & 3 \end{bmatrix} \sim \begin{bmatrix} 1 & 1 & 1 \\ 0 & 1 & -3 \\ 0 & 0 & 12 \end{bmatrix}$$

This Casorati matrix has three pivots and hence is invertible, by the IMT. Hence the set of signals $\{1^k, 2^k, (-2)^k\}$ is linearly independent in $\mathbb{S}$. We know (from the text) that these signals are in the solution space H of a third-order difference equation. By Theorem 18, $\dim H = 3$. Since the three signals are linearly independent, they form a basis for H, by Theorem 13 in Section 5.5.

Warning: Many student papers for Exercise 7 suffer from a lack of precision, often confusing linear independence of the columns of the Casorati matrix with linear independence of the signals in $\mathbb{S}$. There is no need to discuss the columns of the Casorati matrix—just observe that the matrix is invertible. But you must point out that the three *signals* are linearly independent, in order to apply Theorem 13 in Section 5.5 to the vector space H of solutions to the difference equation.

13. The auxiliary equation for $y_{k+2} - y_{k+1} + \frac{2}{9}y_k = 0$ is $r^2 - r + \frac{2}{9} = 0$. By the quadratic formula,

$$r = \frac{1 \pm \sqrt{1 - 8/9}}{2} = \frac{1 \pm 1/3}{2} = \frac{2}{3} \text{ or } \frac{1}{3}$$

So two solutions of the difference equation are $(\frac{2}{3})^k$ and $(\frac{1}{3})^k$. These signals are obviously linearly independent because neither is a multiple of the other. Since the solution space is two-dimensional (Theorem 18), the two signals form a basis for the solution space (Theorem 13).

Study Tip: I think Exercises 7-19 (and 25-28) are good exam questions, because they illustrate how important Theorems 18 and 13 really are. Probably, you do not have to memorize theorem numbers, but your discussion should show that you have two theorems in mind and that you know how to use them. (Check with your instructor.)

19. The auxiliary equation for $y_{k+2} + 4y_{k+1} + y_k = 0$ is $r^2 + 4r + 1 = 0$. By the quadratic formula,

$$r = \frac{-4 \pm \sqrt{16 - 4}}{2} = \frac{-4 \pm 2\sqrt{3}}{2} = -2 \pm \sqrt{3}$$

Two solutions of the difference equation are $(-2 + \sqrt{3})^k$ and $(-2 - \sqrt{3})^k$. They are obviously linearly independent because neither is a multiple of the other. Since the solution space is two-dimensional (Theorem 18), the two signals form a fundamental set of solutions (Theorem 13), and the general solution has the form $c_1(-2 + \sqrt{3})^k + c_2(-2 - \sqrt{3})^k$.

25. To prove that $y_k = k^2$ is a solution of

$$y_{k+2} + 3y_{k+1} - 4y_k = 10k + 7 \qquad (1)$$

show that when k^2, $(k + 1)^2$, and $(k + 2)^2$ are substituted for y_k, y_{k+1}, and y_{k+2}, respectively, the resulting equation is true for all k:

$$\begin{aligned}(k + 2)^2 + 3(k + 1)^2 - 4k^2 &= (k^2 + 4k + 4) + 3(k^2 + 2k + 1) - 4k^2 \\ &= (1 + 3 - 4)k^2 + (4 + 6)k + (4 + 3) \\ &= 10k + 7 \quad \text{for all } k.\end{aligned}$$

So k^2 is a solution of (1). The auxiliary equation for the homogeneous difference equation

$$y_{k+2} + 3y_{k+1} - 4y_k = 0 \quad \text{for all } k \qquad (2)$$

is $r^2 + 3r - 4 = 0$, which factors as $(r - 1)(r + 4) = 0$, so $r = 1, -4$. Thus 1^k and $(-4)^k$ are solutions of (2). The signals are linearly independent (for neither is a multiple of the other), so they form a basis for the two-dimensional solution space. The general solution of (2) is $c_1 \cdot 1^k + c_2(-4)^k$. Add this to a particular solution of (1) and obtain the general solution $k^2 + c_1 + c_2(-4)^k$ of (1).

31. The full explanation is in the text's answer section.

35. For $\{y_k\}$ and $\{z_k\}$ in $\mathbb{S}$, the kth term of $\{y_k\} + \{z_k\}$ is $y_k + z_k$. Hence

$$\begin{aligned} T(\{y_k\} + \{z_k\}) &= (y_{k+2} + z_{k+2}) + a(y_{k+1} + z_{k+1}) + b(y_k + z_k) \\ &= (y_{k+2} + ay_{k+1} + by_k) + (z_{k+2} + az_{k+1} + bz_k) \\ &= T\{y_k\} + T\{z_k\} \end{aligned}$$

For any scalar r, the kth term of $r\{y_k\}$ is ry_k, and so

$$\begin{aligned} T(r\{y_k\}) &= ry_{k+2} + a(ry_{k+1}) + b(ry_k) \\ &= r(y_{k+2} + ay_{k+1} + by_k) = rT\{y_k\} \end{aligned}$$

Thus T has the two properties that define a linear transformation.

Appendix: The Casorati Test

Let $\{\mathbf{y}_1, \ldots, \mathbf{y}_n\}$ be a set of signals in $\mathbb{S}$. For $j = 1, \ldots, n$ and for any k, let $\mathbf{y}_j(k)$ denote the kth entry in the signal $\mathbf{y}_j$ and let

$$C(k) = \begin{bmatrix} \mathbf{y}_1(k) & \cdots & \mathbf{y}_n(k) \\ \mathbf{y}_1(k+1) & \cdots & \mathbf{y}_n(k+1) \\ \vdots & & \vdots \\ \mathbf{y}_1(k+n-1) & \cdots & \mathbf{y}_n(k+n-1) \end{bmatrix} \quad \text{The Casorati matrix}$$

a. If $C(k)$ is invertible for some k, $\{\mathbf{y}_1, \ldots, \mathbf{y}_n\}$ is linearly independent.

b. If $\mathbf{y}_1, \ldots, \mathbf{y}_n$ all satisfy a homogeneous difference equation of order n,

$$y_{k+n} + a_1 y_{k+n-1} + \cdots + a_n y_k = 0 \quad \text{for all } k \qquad (*)$$

(with $a_n \neq 0$), and if the Casorati matrix $C(k)$ is not invertible for some k, then $\{\mathbf{y}_1, \ldots, \mathbf{y}_n\}$ is linearly dependent in $\mathbb{S}$, and for all k, $C(k)$ is not invertible.

Proof. (a) The argument given in the text (page 250) for a set of three signals generalizes immediately to n signals. (b) Suppose that $\mathbf{y}_1, \ldots, \mathbf{y}_n$ are in the set H of solutions of (*) and $C(k_0)$ is not invertible for some k_0. It is readily verified that if $T: H \to \mathbb{R}^n$ is defined by

$$T(\mathbf{y}) = \begin{bmatrix} \mathbf{y}(k_0) \\ \mathbf{y}(k_0+1) \\ \vdots \\ \mathbf{y}(k_0+n-1) \end{bmatrix}$$

then T is a linear transformation. The proof of Theorem 17 is easily modified to show that (*) has a unique solution $\mathbf{y}$ whenever $\mathbf{y}(k_0)$, ..., $\mathbf{y}(k_0+n-1)$ are specified. This means that T is a one-to-one mapping of H onto $\mathbb{R}^n$. Furthermore, the images $T(\mathbf{y}_1)$, ..., $T(\mathbf{y}_n)$ form the columns of the Casorati matrix $C(k_0)$ and hence are linearly dependent, because $C(k_0)$ is not invertible. Since T is one-to-one, $\{\mathbf{y}_1,\ldots,\mathbf{y}_n\}$ is linearly dependent, by Exercise 32 in Section 5.3. In this case, for any k the images $T(\mathbf{y}_1)$, ..., $T(\mathbf{y}_n)$ are linearly dependent, and so $C(k)$ is not invertible. (See Exercise 31 in Section 5.3.)

MATLAB In Exercises 7-16 and 25-28, the polynomial in the auxiliary equation is stored in a row vector $\mathbf{p}$, with coefficients in descending order. For instance, if the auxiliary equation is $r^2 + 6r + 9 = 0$, then $\mathbf{p}$ = [1 6 9].

The MATLAB command **roots(p)** produces a column vector whose entries are the roots of the polynomial described by $\mathbf{p}$.

5.9 APPLICATIONS TO MARKOV CHAINS

This section builds on the population movement example in Section 2.7. You should review that example now. Markov chains are widely used in applications and there is a rich theory connected with them. The simple examples and exercises in this section provide a basic foundation on which you can build later as needed. Two of the examples here will be analyzed from a different point of view in Section 6.2.

KEY IDEAS

A probability vector is a list of nonnegative numbers that sum to one. A Markov chain is a sequence of probability vectors $\{\mathbf{v}_k\}$ that satisfy a difference equation $\mathbf{v}_{k+1} = P\mathbf{v}_k$ $(k = 0,1,\ldots)$ for some stochastic matrix P (whose columns are themselves probability vectors).

The theory of this chapter can be used to show that $P - I$ always has a nontrivial null space when P is a stochastic matrix (Exercise 17). In advanced texts it is shown that the null space of $P - I$ always includes at least one probability vector, which then is a steady-state vector for P, because the equation $(P - I)\mathbf{q} = \mathbf{0}$ is equivalent to $P\mathbf{q} = \mathbf{q}$. (Also, see Exercise 18.)

Our main interest is in a *regular* stochastic matrix P. In this case the steady-state vector is unique, according to Theorem 19. The key to predicting the distant future for a Markov chain associated with such a P is to find the steady-state vector $\mathbf{q}$, since the sequence $\{\mathbf{v}_k\}$ converges to $\mathbf{q}$ no matter what the initial state.

SOLUTIONS TO EXERCISES

1. a. To set up the stochastic matrix P, label the columns N (for news) and M (for music) in some order; use the *same order* for the columns. (Failure to keep the same order is a common source of error in this type of problem.) The data should be arranged so you read *down* a column and then to the right along the rows.

$$\begin{array}{c} \text{From:} \\ \begin{array}{cc} \text{N} & \text{M} \end{array} \end{array} \quad \text{To:}$$

$$\begin{bmatrix} .7 & .6 \\ .3 & .4 \end{bmatrix} \begin{array}{l} \text{News} \\ \text{Music} \end{array}$$

b. You are told that 100% of the listeners are listening to the news at 8:15 a.m., so start the Markov chain then, with $\mathbf{v}_0 = \begin{bmatrix} 1 \\ 0 \end{bmatrix}$.

c. There are two breaks between 8:15 and 9:25, so you need $\mathbf{v}_2$.

$$\mathbf{v}_1 = P\mathbf{v}_0 = \begin{bmatrix} .7 & .6 \\ .3 & .4 \end{bmatrix}\begin{bmatrix} 1 \\ 0 \end{bmatrix} = \begin{bmatrix} .7 \\ .3 \end{bmatrix}$$

$$\mathbf{v}_2 = P\mathbf{v}_1 = \begin{bmatrix} .7 & .6 \\ .3 & .4 \end{bmatrix}\begin{bmatrix} .7 \\ .3 \end{bmatrix} = \begin{bmatrix} .67 \\ .33 \end{bmatrix}$$

The entries in $\mathbf{v}_2$ show that after two station breaks, 67% of the audience is listening to the news and 33% is listening to music.

Study Tip: When you need to know a typical probability vector $P\mathbf{v}$, be sure to *compute all* of the entries. Then check your work by verifying that the entries sum to 1. (This trick has caught numerous errors of mine!)

7. To find the steady state vector for a regular stochastic matrix P:
   (i) set up the matrix $P - I$;
   (ii) find the general solution of $(P - I)\mathbf{x} = \mathbf{0}$;
   (iii) choose a basis vector for $\text{Nul}(P - I)$ whose entries sum to 1.

$$P = \begin{bmatrix} .7 & 1 & .1 \\ .2 & .8 & .2 \\ .1 & .1 & .7 \end{bmatrix}, \quad P - I = \begin{bmatrix} .7 & 1 & .1 \\ .2 & .8 & .2 \\ .1 & .1 & .7 \end{bmatrix} - \begin{bmatrix} 1 & 0 & 0 \\ 0 & 1 & 0 \\ 0 & 0 & 1 \end{bmatrix} = \begin{bmatrix} -.3 & .1 & .1 \\ .2 & -.2 & .2 \\ .1 & .1 & -.3 \end{bmatrix}$$

Solve $(P - I)\mathbf{x} = \mathbf{0}$:

$$\begin{bmatrix} -.3 & .1 & .1 & 0 \\ .2 & -.2 & .2 & 0 \\ .1 & .1 & -.3 & 0 \end{bmatrix} \sim \begin{bmatrix} 1 & 1 & -3 & 0 \\ 2 & -2 & 2 & 0 \\ -3 & 1 & 1 & 0 \end{bmatrix}$$

Interchange rows 1 and 3
Scale every row by 10

$$\sim \begin{bmatrix} 1 & 1 & -3 & 0 \\ 0 & -4 & 8 & 0 \\ 0 & 4 & -8 & 0 \end{bmatrix} \sim \begin{bmatrix} 1 & 1 & -3 & 0 \\ 0 & 1 & -2 & 0 \\ 0 & 0 & 0 & 0 \end{bmatrix}$$

Add row 2 to row 3
Scale row 2 by -1/4

$$\sim \begin{bmatrix} 1 & 0 & -1 & 0 \\ 0 & 1 & -2 & 0 \\ 0 & 0 & 0 & 0 \end{bmatrix}, \quad \begin{aligned} x_1 - x_3 &= 0 \\ x_2 - 2x_3 &= 0 \\ x_3 &\text{ is free} \end{aligned}; \quad \begin{bmatrix} x_1 \\ x_2 \\ x_3 \end{bmatrix} = \begin{bmatrix} x_3 \\ 2x_3 \\ x_3 \end{bmatrix} = x_3\begin{bmatrix} 1 \\ 2 \\ 1 \end{bmatrix}$$

The entries in $\begin{bmatrix} 1 \\ 2 \\ 1 \end{bmatrix}$ sum to 4, so $\mathbf{q} = \frac{1}{4}\begin{bmatrix} 1 \\ 2 \\ 1 \end{bmatrix} = \begin{bmatrix} 1/4 \\ 1/2 \\ 1/4 \end{bmatrix}$ or $\begin{bmatrix} .25 \\ .50 \\ .25 \end{bmatrix}$.

Study Tip: Notice that the column sums are all zero for the matrix $P - I$ of Exercise 7. This always happens (see Exercise 17), and so you have a fast way to check your arithmetic for the entries in $P - I$.

13. From Exercise 3, $P = \begin{bmatrix} .50 & .25 & .25 \\ .25 & .50 & .25 \\ .25 & .25 & .50 \end{bmatrix}$. So $P - I = \begin{bmatrix} -.50 & .25 & .25 \\ .25 & -.50 & .25 \\ .25 & .25 & -.50 \end{bmatrix}$.

Solve $(P - I)\mathbf{x} = \mathbf{0}$:

$$\begin{bmatrix} -.50 & .25 & .25 & 0 \\ .25 & -.50 & .25 & 0 \\ .25 & .25 & -.50 & 0 \end{bmatrix} \sim \begin{bmatrix} 1 & 1 & -2 & 0 \\ 1 & -2 & 1 & 0 \\ -2 & 1 & 1 & 0 \end{bmatrix}$$

Scale rows by 4
Interchange rows 1 & 3

$$\sim \begin{bmatrix} 1 & 1 & -2 & 0 \\ 0 & -3 & 3 & 0 \\ 0 & 3 & -3 & 0 \end{bmatrix} \sim \begin{bmatrix} 1 & 1 & -2 & 0 \\ 0 & 1 & -1 & 0 \\ 0 & 0 & 0 & 0 \end{bmatrix} \sim \begin{bmatrix} 1 & 0 & -1 & 0 \\ 0 & 1 & -1 & 0 \\ 0 & 0 & 0 & 0 \end{bmatrix} \quad \begin{aligned} x_1 &= x_3 \\ x_2 &= x_3 \\ x_3 &\text{ is free} \end{aligned}$$

A basis for Nul$(P - I)$ is $\left\{\begin{bmatrix} 1 \\ 1 \\ 1 \end{bmatrix}\right\}$; the steady-state vector is $\mathbf{q} = \begin{bmatrix} 1/3 \\ 1/3 \\ 1/3 \end{bmatrix}$.

So each food will be preferred equally.

19. a. The product $S\mathbf{v}$ equals the sum of the entries in $\mathbf{v}$. Thus, by definition, $\mathbf{v}$ is a probability vector if and only if its entries are nonnegative and $S\mathbf{v} = 1$.

b. Let $P = [\mathbf{p}_1 \ \ \mathbf{p}_2 \ \ \cdots \ \ \mathbf{p}_n]$, where the $\mathbf{p}_i$ are probability vectors. By matrix multiplication and part (a),

$$SP = [S\mathbf{p}_1 \ \ S\mathbf{p}_2 \ \ \cdots \ \ S\mathbf{p}_n] = [1 \ \ 1 \ \ \cdots \ \ 1] = S$$

c. By part (b), $S(P\mathbf{v}) = (SP)\mathbf{v} = S\mathbf{v} = 1$. The entries in $P\mathbf{v}$ are obviously nonnegative, because P and $\mathbf{v}$ have only nonnegative entries. By (a), the condition $S(P\mathbf{v}) = 1$ shows that $P\mathbf{v}$ is a probability vector.

Appendix: Hand Computations With a Calculator

Without question, the presence of so many decimals in this section makes one want to use a computer or a supercalculator that has built-in matrix operations. However, if neither is available, a scientific calculator should do the "chain" calculations needed for the entries in $P\mathbf{v}$.

On a calculator that has "parentheses" keys, the quantity

$$3\times4 + 5\times6 + 7\times8$$

is found by pressing the following keys:

[3] [×] [4] [+] [5] [×] [6] [+] [7] [×] [8] [=]

The displayed answer should be 98. On a calculator that uses RPN (reverse Polish notation), such as the Hewlett-Packard models, press

[3] [Enter] [4] [×] [5] [Enter] [6] [×] [+] [7] [Enter] [8] [×] [+]

- [3] [Enter] [4] [×]: computes 3×4
- [5] [Enter] [6] [×]: computes 5×6
- [+]: adds two products
- [7] [Enter] [8] [×]: computes 7×8
- [+]: adds 3rd product to the others

MATLAB The MATLAB note for Section 2.7 contains information that is useful for homework here.

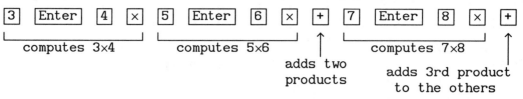

CHAAPTER 5 GLOSSARY CHECKLIST

Check your knowledge by attempting to write definitions of the terms below. Then compare your work with the definitions given in the text's Glossary. Ask your instructor which definitions, if any, might appear on a test.

auxiliary equation: A polynomial equation in a variable r, created from ...

basis (for a subspace H): A set $\mathcal{B} = \{\mathbf{v}_1, \ldots, \mathbf{v}_p\}$ in V such that: ....

$\mathcal{B}$-coordinates of x: *See* coordinates of **x** relative to the basis $\mathcal{B}$.

change-of-coordinates matrix (from a basis $\mathcal{B}$ to a basis $\mathcal{C}$): A matrix $\underset{\mathcal{C}\leftarrow\mathcal{B}}{P}$ that transforms ..., namely, (equation) ....

column space (of an $m \times n$ matrix A): The set Col A of .... In set notation, Col $A = \{ \quad : \qquad \}$.

controllable (pair of matrices): A matrix pair (A, B) where A is $n \times n$, B has n rows, and ....

coordinate mapping (determined by an ordered basis $\mathcal{B}$ in a vector space V): a mapping that associates to each ....

coordinates of x relative to the basis $\mathcal{B} = \{\mathbf{b}_1, \ldots, \mathbf{b}_n\}$: ....

coordinate vector of x relative to $\mathcal{B}$: The vector $[\mathbf{x}]_{\mathcal{B}}$ whose entries ....

dimension (of a vector space V): The number ....

explicit description (of a subspace W of $\mathbb{R}^n$): A parametric representation of W as the set of ....

finite-dimensional (vector space): A vector space that is ....

full rank (matrix): An $m \times n$ matrix whose rank is ....

fundamental set of solutions: A ... for the set of solutions of ....

fundamental subspaces (of A): The ... of A, and ....

implicit description (of a subspace W of $\mathbb{R}^n$): A set of one or more ....

infinite-dimensional (vector space): A nonzero vector space V that ....

isomorphism: A ... mapping from one vector space ....

kernel (of a linear transformation $T: V \to W$): The set of ... such that ....

linear combination: Any sum ....

linear dependence relation: A ... equation where ....

linear filter: A ... equation used to transform ....

linearly dependent (vectors): A set $\{\mathbf{v}_1, \ldots, \mathbf{v}_p\}$ with the property that ....

linearly independent (vectors): A set $\{\mathbf{v}_1, \ldots, \mathbf{v}_p\}$ with the property ....

linear transformation T (from a vector space V into a vector space W): A rule $T: V \to W$ that to each vector $\mathbf{x}$ in V assigns a unique vector $T(\mathbf{x})$ in W, such that: ....

Markov chain: A sequence of ... vectors $\mathbf{v}_0$, $\mathbf{v}_1$, $\mathbf{v}_2, \ldots$, together with a ... matrix P such that ....

maximal linearly independent set (in V): A linearly independent set $\mathcal{B}$ in V such that if ..., then ...

minimal spanning set (for a subspace H): A set $\mathcal{B}$ that spans H and has the property that if ..., then ....

null space (of an $m \times n$ matrix A): The set Nul A of all .... In set notation, Nul $A = \{ \quad : \quad \}$.

probability vector: A vector in $\mathbb{R}^n$ whose entries ....

proper subspace: Any subspace of a vector space V ....

range (of a linear transformation $T: V \longrightarrow W$): The set of all vectors ....

rank (of a matrix A): ....

regular stochastic matrix: A ... matrix P such that ....

row space (of a matrix A): The set Row A of all ...; also denoted by ....

signal (or **discrete-time signal**): ....

Span $\{\mathbf{v}_1, \ldots, \mathbf{v}_p\}$: The set of .... Also, the ... *spanned* ....

spanning set (for a subspace H): Any set $\{\mathbf{v}_1, \ldots, \mathbf{v}_p\}$ ... such that ....

standard basis: The basis ... for $\mathbb{R}^n$ consisting of ..., or the basis ... for $\mathbb{P}_n$.

state vector: A ... vector. In general, a vector that ... , often in connection with a difference equation ....

steady-state vector (for a stochastic matrix P): A ... vector $\mathbf{v}$ such ....

stochastic matrix: A ... matrix ....

submatrix (of A): Any matrix obtained by ....

subspace: A subset H of some vector space V such that ....

vector space: A set of objects, called vectors, on which ....

zero subspace: The subspace ... consisting of ....

6

Eigenvalues and Eigenvectors

6.1 EIGENVECTORS AND EIGENVALUES

This section introduces eigenvectors and eigenvalues. A hint about the connection with dynamical systems appears at the end of the section.

KEY IDEAS

In words, a nonzero vector $\mathbf{v}$ is an eigenvector of a matrix A if and only if the transformed vector $A\mathbf{x}$ points in the same or opposite direction of $\mathbf{v}$.

Notice that while an eigenvalue λ might be zero, an eigenvector is never zero (by definition). An eigenspace contains eigenvectors together with the zero vector.

The two equations $A\mathbf{x} = \lambda\mathbf{x}$ and $(A - \lambda I)\mathbf{x} = \mathbf{0}$ are equivalent. See Example 3. The first equation is useful for understanding what eigenvalues and eigenvectors are, and it shows the geometric effect of the linear transformation $\mathbf{x} \mapsto A\mathbf{x}$ on an eigenvector. The second equation shows that the eigenspace is a subspace (because it is the null space of the matrix $A - \lambda I$), and the equation is used to find a basis for the eigenspace, when λ is a known eigenvalue. The second equation will be used again in Section 6.2 for another purpose.

SOLUTIONS TO EXERCISES

1. The number 2 is an eigenvalue of A if and only if the equation $A\mathbf{x} = 2\mathbf{x}$ has a nontrivial solution. This equation is equivalent to $(A - 2I)\mathbf{x} = \mathbf{0}$. Compute

$$A - 2I = \begin{bmatrix} 3 & 2 \\ 3 & 8 \end{bmatrix} - \begin{bmatrix} 2 & 0 \\ 0 & 2 \end{bmatrix} = \begin{bmatrix} 1 & 2 \\ 3 & 6 \end{bmatrix}$$

The columns of A are obviously linearly dependent, so $(A - 2I)\mathbf{x} = \mathbf{0}$ has a nontrivial solution, and so 2 is an eigenvalue of A.

7. Proceed as in Exercise 1:

$$A - 4I = \begin{bmatrix} 3 & 0 & -1 \\ 2 & 3 & 1 \\ -3 & 4 & 5 \end{bmatrix} - \begin{bmatrix} 4 & 0 & 0 \\ 0 & 4 & 0 \\ 0 & 0 & 4 \end{bmatrix} = \begin{bmatrix} -1 & 0 & -1 \\ 2 & -1 & 1 \\ -3 & 4 & 1 \end{bmatrix}$$

We need to know whether $A - 4I$ is invertible. This could be checked in several ways, but since we are asked for an eigenvector, in the event that one exists, the best strategy is to row reduce the augmented matrix for $(A - 4I)\mathbf{x} = \mathbf{0}$:

$$\begin{bmatrix} -1 & 0 & -1 & 0 \\ 2 & -1 & 1 & 0 \\ -3 & 4 & 1 & 0 \end{bmatrix} \sim \begin{bmatrix} -1 & 0 & -1 & 0 \\ 0 & -1 & -1 & 0 \\ 0 & 4 & 4 & 0 \end{bmatrix} \sim \begin{bmatrix} -1 & 0 & -1 & 0 \\ 0 & -1 & -1 & 0 \\ 0 & 0 & 0 & 0 \end{bmatrix}$$

Now it is clear that 4 is an eigenvalue of A [because $(A - 4I)\mathbf{x} = \mathbf{0}$ has a nontrivial solution]. The coordinates of an eigenvector satisfy $-x_1 - x_3 = 0$ and $-x_2 - x_3 = 0$. The general solution is not requested, so take any nonzero value for x_3 to produce an eigenvector. If $x_3 = 1$, then $\mathbf{x} = (-1, -1, 1)$.

Checkpoint 1: The answer in the text is different, namely, $(1, 1, -1)$. Can it also be correct?

Helpful Hint: Suppose you think that 4 is an eigenvalue of a matrix, as in Exercise 7, and you row reduce the augmented matrix for $(A - 4I)\mathbf{x} = \mathbf{0}$. If you discover that there are no free variables, then there are only two possibilities: (1) 4 is *not* an eigenvalue of A, or (2) you have made an error.

13. <u>For $\lambda = 1$:</u>

$$A - 1I = \begin{bmatrix} 4 & 0 & 1 \\ -2 & 1 & 0 \\ -2 & 0 & 1 \end{bmatrix} - \begin{bmatrix} 1 & 0 & 0 \\ 0 & 1 & 0 \\ 0 & 0 & 1 \end{bmatrix} = \begin{bmatrix} 3 & 0 & 1 \\ -2 & 0 & 0 \\ -2 & 0 & 0 \end{bmatrix}$$

The equations for $(A - I)\mathbf{x} = \mathbf{0}$ are easy to solve: $\begin{cases} 3x_1 + x_3 = 0 \\ -2x_1 = 0 \end{cases}$

Row operations hardly seem necessary. Obviously x_1 is zero, and hence x_3 is also zero. There are three variables, so x_2 is free. The general solution of $(A - I)\mathbf{x} = \mathbf{0}$ is $x_2\mathbf{e}_2$, where $\mathbf{e}_2 = \begin{bmatrix} 0 \\ 1 \\ 0 \end{bmatrix}$, and so $\mathbf{e}_2$ provides a basis for the eigenspace.

<u>For $\lambda = 2$:</u>

$$A - 2I = \begin{bmatrix} 4 & 0 & 1 \\ -2 & 1 & 0 \\ -2 & 0 & 1 \end{bmatrix} - \begin{bmatrix} 2 & 0 & 0 \\ 0 & 2 & 0 \\ 0 & 0 & 2 \end{bmatrix} = \begin{bmatrix} 2 & 0 & 1 \\ -2 & -1 & 0 \\ -2 & 0 & -1 \end{bmatrix}$$

$$[(A - 2I)\ \mathbf{0}] = \begin{bmatrix} 2 & 0 & 1 & 0 \\ -2 & -1 & 0 & 0 \\ -2 & 0 & -1 & 0 \end{bmatrix} \sim \begin{bmatrix} 2 & 0 & 1 & 0 \\ 0 & -1 & 1 & 0 \\ 0 & 0 & 0 & 0 \end{bmatrix} \sim \begin{bmatrix} 1 & 0 & 1/2 & 0 \\ 0 & 1 & -1 & 0 \\ 0 & 0 & 0 & 0 \end{bmatrix}$$

So $x_1 = -(1/2)x_3$, $x_2 = x_3$, with x_3 free. The general solution of $(A - 2I)\mathbf{x} = \mathbf{0}$ is $x_3\begin{bmatrix} -1/2 \\ 1 \\ 1 \end{bmatrix}$. A nice basis for the eigenspace is $\begin{bmatrix} -1 \\ 2 \\ 2 \end{bmatrix}$.

<u>For $\lambda = 3$:</u>

$$A - 3I = \begin{bmatrix} 4 & 0 & 1 \\ -2 & 1 & 0 \\ -2 & 0 & 1 \end{bmatrix} - \begin{bmatrix} 3 & 0 & 0 \\ 0 & 3 & 0 \\ 0 & 0 & 3 \end{bmatrix} = \begin{bmatrix} 1 & 0 & 1 \\ -2 & -2 & 0 \\ -2 & 0 & -2 \end{bmatrix}$$

$$[(A - 3I)\ \mathbf{0}] = \begin{bmatrix} 1 & 0 & 1 & 0 \\ -2 & -2 & 0 & 0 \\ -2 & 0 & -2 & 0 \end{bmatrix} \sim \begin{bmatrix} 1 & 0 & 1 & 0 \\ 0 & -2 & 2 & 0 \\ 0 & 0 & 0 & 0 \end{bmatrix} \sim \begin{bmatrix} 1 & 0 & 1 & 0 \\ 0 & 1 & -1 & 0 \\ 0 & 0 & 0 & 0 \end{bmatrix}$$

So $x_1 = -x_3$, $x_2 = x_3$, with x_3 free. A basis for the eigenspace is $\begin{bmatrix} -1 \\ 1 \\ 1 \end{bmatrix}$.

Study Tip: The text's answer to Exercise 15 is likely to be the same as yours. But there are many answers. What should you do if your vectors differ from those in the answer key? Example 2 gives the answer. Whenever you compute an $\mathbf{x}$ that you think is an eigenvector of A, you can check this simply by computing $A\mathbf{x}$. There is a little more to do in Exercise 15, however. The answer shows a basis of two eigenvectors, which means that the eigenspace is two-dimensional. So your answer must consist of two linearly independent eigenvectors. You can check that they are indeed eigenvectors, and then their linear independence can be checked by inspection.

19. Since the matrix $\begin{bmatrix} 0 & 0 & 0 \\ 0 & 2 & 5 \\ 0 & 0 & -1 \end{bmatrix}$ is triangular, its eigenvalues lie on the diagonal, and are 0, 2, -1.

Checkpoint 2: What kind of matrix has 0 as an eigenvalue?

23. If a 2×2 matrix A had three distinct eigenvalues, then by Theorem 2 there would correspond three linearly independent eigenvectors (one for each eigenvalue). This is impossible because the vectors all belong to a two-dimensional vector space in which any set of three vectors is linearly dependent. See Theorem 6 in Section 2.4. In general, if an $n \times n$ matrix has p distinct eigenvalues, then by Theorem 2 there would be a linearly independent set of p eigenvectors (one for each eigenvalue). Since these vectors belong to an n-dimensional vector space, p cannot exceed n.

25. Let $\mathbf{x}$ be a nonzero vector such that $A\mathbf{x} = \lambda\mathbf{x}$. Then $A^{-1}A\mathbf{x} = A^{-1}(\lambda\mathbf{x})$, and $\mathbf{x} = \lambda(A^{-1}\mathbf{x})$. Since $\mathbf{x} \neq \mathbf{0}$ (and since A is invertible), λ cannot be zero. Then $\lambda^{-1}\mathbf{x} = A^{-1}\mathbf{x}$, which shows that λ^{-1} is an eigenvalue of A^{-1}.

Note: The relation between the eigenvalues of A and A^{-1} is important in the so-called *inverse power method* for estimating an eigenvalue of a matrix. (See Section 6.7.)

27. For any λ, $(A - \lambda I)^T = A^T - (\lambda I)^T = A^T - \lambda I$. Since $(A - \lambda I)^T$ is invertible if and only if $A - \lambda I$ is invertible (by Theorem 6 in Section 3.2), we conclude that $A^T - \lambda I$ is *not* invertible if and only if $A - \lambda I$ is *not* invertible. That is, λ is an eigenvalue of A^T if and only if λ is an eigenvalue of A.

31. The solution is given in the text. This exercise is important because it sets the stage for work later in the chapter. You ought to spend at least a little time on Exercise 32, too, even if that is not assigned.

Here is a more sophisticated solution of Exercise 31(b). The discussion at the end of the section shows that $\lambda^k\mathbf{u}$ and $\mu^k\mathbf{v}$ are both solutions of the difference equation $\mathbf{x}_{k+1} = A\mathbf{x}_k$, and Exercise 31(b) asks us to show that any linear combination of these solutions is again a solution. This fact follows from the fact that the set of all solutions is a subspace H of the vector space V of all sequences of vectors in $\mathbb{R}^n$. Finally, H is a subspace because it is the kernel of the linear transformation that takes the sequence $\{\mathbf{x}_k\}$ into the sequence $\{\mathbf{x}_{k+1} - A\mathbf{x}_k\}$. (It is not difficult to verify that this transformation is linear.)

Answers to Checkpoints:

1. Of course, because the answer in the text is a multiple of the eigenvector found in the solution to Exercise 7, and any nonzero multiple of an eigenvector is another eigenvector. (Remember, the eigenspace is a *vector space*, so it is closed under scalar multiplication.)

2. A matrix has 0 as an eigenvalue if and only if the matrix is square and not invertible, because the equation $A\mathbf{x} = 0\mathbf{x}$ is the same as the equation $A\mathbf{x} = \mathbf{0}$. See the discussion following Example 5.

MATLAB: The linear algebra Toolbox has a command that will simplify your homework by automatically producing a basis for an eigenspace. For example, if A is a 3×3 matrix with an eigenvalue 7, then the commands

```
C = A - 7*eye(3)
B = nulbasis(C)
```

or, simply, **B = nulbasis(A - 7*eye(3))**, produce a matrix B whose columns form a basis for the eigenspace for A corresponding to $\lambda = 7$. In general, **eye(k)** is a $k \times k$ identity matrix, and **nulbasis(C)** is a matrix whose columns form a basis for Nul C.

Remarks:

1. If the numbers in the basis matrix B are messy, try the command

   rat(B,'s')

   which will display the answer to the **nulbasis** command as a matrix with rational entries, if possible.
2. Although MATLAB has more powerful commands for eigenvector calculations, you should not use them yet, because you need to learn basic concepts about eigenvectors and eigenvalues.

6.2 THE CHARACTERISTIC EQUATION

There are several equivalent definitions of the determinant of a matrix. The definition here in terms of the pivots in an echelon form has the advantage that it is easy to state and understand, and it provides a fairly rapid way to compute a determinant.

KEY IDEAS

When A is 3×3, the geometric interpretation of det A as a volume explains why det $A = 0$ if and only if A is not invertible:

The determinant of A is zero.
⇔ The parallelepiped determined by the columns of A has zero volume.
⇔ One column of A is in the subspace spanned by the other columns.
⇔ The columns of A are linearly dependent.
⇔ The matrix A is not invertible.

If A is $n \times n$, then $\det(A - \lambda I) = 0$ if and only if $A - \lambda I$ is not invertible, and this happens if and only if λ is an eigenvalue of A.

Exercises 1—14 are designed only to provide some basic familiarity with characteristic polynomials. The main use of $\det(A - \lambda I)$ is as a tool for *studying* eigenvalues rather than computing them.

Sometimes the characteristic polynomial is defined as $\det(\lambda I - A)$. A property of determinants implies that $\det(\lambda I - A) = (-1)^n \det(A - \lambda I)$, when A is $n \times n$, so the two polynomials are either the same (when n is even) or they are negatives of one another. The use of $\det(A - \lambda I)$ tends to make hand calculations easier and less prone to copying errors.

SOLUTIONS TO EXERCISES

1. $A = \begin{bmatrix} 2 & 7 \\ 7 & 2 \end{bmatrix}$, $A - \lambda I = \begin{bmatrix} 2 & 7 \\ 7 & 2 \end{bmatrix} - \begin{bmatrix} \lambda & 0 \\ 0 & \lambda \end{bmatrix} = \begin{bmatrix} 2-\lambda & 7 \\ 7 & 2-\lambda \end{bmatrix}$, the characteristic polynomial is

$$\det(A - \lambda I) = (2 - \lambda)^2 - 7^2 = 4 - 4\lambda + \lambda^2 - 49 = \lambda^2 - 4\lambda - 45$$

In factored form, the characteristic equation is $(\lambda - 9)(\lambda + 5) = 0$, so the eigenvalues of A are 9 and -5.

Warning: Don't row reduce a matrix A to find its eigenvalues. Row reduction preseves the null space of A but not the eigenvalues of A.

7. $A = \begin{bmatrix} 5 & 3 \\ -4 & 4 \end{bmatrix}$, $A - \lambda I = \begin{bmatrix} 5-\lambda & 3 \\ -4 & 4-\lambda \end{bmatrix}$, the characteristic polynomial is

$$\det(A - \lambda I) = (5 - \lambda)(4 - \lambda) - (3)(-4) = 20 - 9\lambda + \lambda^2 + 12$$
$$= \lambda^2 - 9\lambda + 32$$

The characteristic polynomial does not factor easily, but the quadratic formula provides the solutions of $\lambda^2 - 9\lambda + 32 = 0$.

$$\lambda = \frac{+9 \pm \sqrt{81 - 4(32)}}{2}$$

These values for λ are not real numbers, so A has no real eigenvalues. There is no vector $\mathbf{x}$ in $\mathbb{R}^2$ such that $A\mathbf{x} = \lambda\mathbf{x}$ for such a λ. (For any $\mathbf{x}$ in $\mathbb{R}^2$, the vector $A\mathbf{x}$ has only real entries and thus could not equal a complex multiple of $\mathbf{x}$.)

Study Tip: If you are asked to work some of Exercises 9-13, you may be tested on them. This is one way of finding out if you know what the characteristic polynomial is and how it is connected with eigenvalues. Also, you can show that you know some elementary properties of determinants.

13. $A - \lambda I = \begin{bmatrix} 6-\lambda & -2 & 0 \\ -2 & 9-\lambda & 0 \\ 5 & 8 & 3-\lambda \end{bmatrix}$

The method using the special 3×3 formula will produce the characteristic polynomial $-\lambda^3 + 18\lambda^2 - 95\lambda + 150$. Factoring such a polynomial to find the eigenvalues requires a little experience. (See the appendix at the end of the exercise solutions.) However, if you use a cofactor expansion down the third column (see Section 4.1), you immediately obtain

$$\det(A - \lambda I) = (3 - \lambda)\cdot\det\begin{bmatrix} 6-\lambda & -2 \\ -2 & 9-\lambda \end{bmatrix}$$

$$= (3 - \lambda)[(6-\lambda)(9-\lambda) - 4]$$

$$= (3 - \lambda)(\lambda^2 - 15\lambda + 50)$$

The characteristic polynomial is already partially factored, and the remaining quadratic factor is itself easily factored. The factored characteristic polynomial is $(3 - \lambda)(\lambda - 10)(\lambda - 5)$ or, equivalently, $-(\lambda - 3)(\lambda - 5)(\lambda - 10)$.

Note: The solution of Exercise 14 is similar to that of Exercise 13. Exercises 11 and 12 are even nicer. They have the property that if a cofactor expansion is chosen along a column or row that contains two zeros, then the characteristic polynomial appears in a partially factored form.

19. Since the equation $\det(A - \lambda I) = (\lambda_1 - \lambda)(\lambda_2 - \lambda)\cdots(\lambda_n - \lambda)$ holds for all λ, set $\lambda = 0$ and conclude that $\det A = \lambda_1\lambda_2\cdots\lambda_n$.

21. If $A = QR$, with Q invertible, and if $A_1 = RQ$, then $QA_1Q^{-1} = Q(RQ)Q^{-1} = QR(QQ^{-1}) = AI = A$. That is, $(Q^{-1})^{-1}A_1Q^{-1} = A$, so A_1 is similar to A.

Another proof: Since Q is invertible, $A_1 = RQ = Q^{-1}QRQ = Q^{-1}AQ = Q^{-1}A(Q^{-1})^{-1}$. Thus A_1 has the form PAP^{-1}, and so A_1 is similar to A.

25. a. $A\mathbf{u}_1 = \mathbf{u}_1$, $A\mathbf{u}_2 = .5\mathbf{u}_2$, $A\mathbf{u}_3 = .2\mathbf{u}_3$ (This also shows that the eigenvalues of A are 1, .5, and .2.)

b. $\{\mathbf{u}_1, \mathbf{u}_2, \mathbf{u}_3\}$ is linearly independent because the eigenvectors correspond to distinct eigenvalues (Theorem 2). Since there are 3 vectors in the set, the set is a basis for $\mathbb{R}^3$. So there exist (unique) constants such that

$$\mathbf{v}_0 = c_1\mathbf{u}_1 + c_2\mathbf{u}_2 + c_3\mathbf{u}_3$$

Then

$$\mathbf{w}^T\mathbf{v}_0 = c_1\mathbf{w}^T\mathbf{u}_1 + c_2\mathbf{w}^T\mathbf{u}_2 + c_3\mathbf{w}^T\mathbf{u}_3 \qquad (*)$$

Since $\mathbf{v}_0$ and $\mathbf{u}_1$ are probability vectors and since the entries in $\mathbf{u}_2$ and in $\mathbf{u}_3$ each sum to 0, (*) shows that $1 = c_1$.

c. By (b),

$$\mathbf{v}_0 = \mathbf{u}_1 + c_2\mathbf{u}_2 + c_3\mathbf{u}_3$$

Using (a),

$$\mathbf{v}_k = A^k\mathbf{v}_0 = A^k\mathbf{u}_1 + c_2A^k\mathbf{u}_2 + c_3A^k\mathbf{u}_3 = \mathbf{u}_1 + c_2(.5)^k\mathbf{u}_2 + c_3(.2)^k\mathbf{u}_3$$

$$\to \mathbf{u}_1 \text{ as } k \to \infty$$

Appendix: Factoring a Polynomial

In general it is difficult to factor a polynomial of degree 3 or higher (unless you have one of several powerful computer programs available). Fortunately, text-book examples and exercises tend to have small integer solutions. The following observation is helpful.

> *Let* $p(\lambda)$ *be a polynomial with integer coefficients. If* $p(c) = 0$ *for some integer* c*, then* $\lambda - c$ *is a factor of* $p(\lambda)$ *and* c *is a divisor of the constant term of* $p(\lambda)$.

EXAMPLE Find the eigenvalues of the matrix A whose characteristic polynomial is $p(\lambda) = -\lambda^3 + 18\lambda^2 - 96\lambda + 160$.

Solution By the observation above, any integer eigenvalue of A must be a divisor of the constant term 160 in the characteristic polynomial. There are twenty-four such divisors: 1, 2, 4, 5, 8, 10, 16, 20, 32, 40, 80, and 160, together with the negatives of these numbers. Start computing $p(c)$ as c ranges over these numbers. Eventually, one finds that $p(4) = 0$, which means that 4 is an eigenvalue of A and $\lambda - 4$ is a factor of $p(\lambda)$. The long division on the next page shows that $p(\lambda) = (\lambda - 4)(-\lambda^2 + 14\lambda - 40)$. The quadratic polynomial factors easily, and the characteristic equation is $(\lambda - 4)(\lambda - 4)(10 - \lambda) = 0$. The eigenvalues of A are 4 (with multiplicity two) and 10. □

$$
\begin{array}{r|l}
 & -\lambda^2 + 14\lambda - 40 \\
\cline{2-2}
\lambda - 4 & -\lambda^3 + 18\lambda^2 - 96\lambda + 160 \\
 & -\lambda^3 + 4\lambda^2 \\
\cline{2-2}
 & \qquad 14\lambda^2 - 96\lambda \\
 & \qquad 14\lambda^2 - 56\lambda \\
\cline{2-2}
 & \qquad\qquad -40\lambda + 160 \\
 & \qquad\qquad -40\lambda + 160 \\
\cline{2-2}
\end{array}
$$

(Sample long division)

MATLAB: You can use the MATLAB command **poly(A)** to check your answers in Exercises 9—14. Note that if A is $n \times n$, **poly(A)** lists the coefficients of the characteristic polynomial of A, in order of decreasing powers of λ, beginning with λ^n. If the polynomial is of odd degree, the coefficients are multiplied by -1, to make $+1$ the coefficient of λ^n. This corresponds to using $\det(\lambda I - A)$.

6.3 DIAGONALIZATION

The factorization $A = PDP^{-1}$ is used to compute powers of A, decouple a dynamical system in Section 6.6, and study symmetric matrices and quadratic forms in Chapter 8.

KEY IDEAS

Example 3 gives the algorithm for diagonalizing a matrix A. After you construct P and D, check your calculations:

1. Compute AP and PD, and check that $AP = PD$.
2. Make sure the columns of P are linearly independent. Use Theorem 7 to save time. You only have to verify that for each eigenvalue, the corresponding eigenvectors are linearly independent. That's easy if the dimension of the eigenspace is 2 or 1.

The key equation in this section is $AP = PD$. It will help you to keep the order of the factors correct when you write $A = PDP^{-1}$, and it also leads immediately to $P^{-1}AP = D$, which you may need occasionally. Possible test question: If $AP = PD$, explain why the first column of P is an eigenvector of A (if the column is nonzero). (Study the proof of Theorem 5.)

Warning: Do not confuse the property of being diagonalizable with the property of being invertible. They are not connected. The matrix in Example 5 is diagonalizable, but it is not invertible because 0 is an eigenvalue. The matrix $\begin{bmatrix} 1 & -2 \\ 0 & 1 \end{bmatrix}$ is invertible, but it is not diagonalizable because the eigenspace for $\lambda = 1$ is only one-dimensional.

SOLUTIONS TO EXERCISES

1. $P = \begin{bmatrix} 5 & 7 \\ 2 & 3 \end{bmatrix}$, $D = \begin{bmatrix} 2 & 0 \\ 0 & 1 \end{bmatrix}$, $A = PDP^{-1}$, and $A^4 = PD^4P^{-1}$. Next, compute

$P^{-1} = \frac{1}{1}\begin{bmatrix} 3 & -7 \\ -2 & 5 \end{bmatrix}$, $D^4 = \begin{bmatrix} 2^4 & 0 \\ 0 & 1 \end{bmatrix} = \begin{bmatrix} 16 & 0 \\ 0 & 1 \end{bmatrix}$. Putting this together,

$$A^4 = \begin{bmatrix} 5 & 7 \\ 2 & 3 \end{bmatrix}\begin{bmatrix} 16 & 0 \\ 0 & 1 \end{bmatrix}\begin{bmatrix} 3 & -7 \\ -2 & 5 \end{bmatrix} = \begin{bmatrix} 80 & 7 \\ 32 & 3 \end{bmatrix}\begin{bmatrix} 3 & -7 \\ -2 & 5 \end{bmatrix} = \begin{bmatrix} 226 & -525 \\ 90 & -209 \end{bmatrix}$$

7. $A = \begin{bmatrix} 1 & 0 \\ 6 & -1 \end{bmatrix}$. The eigenvalues are obviously ± 1 (since A is triangular).

<u>For $\lambda = 1$:</u> $A - 1I = \begin{bmatrix} 0 & 0 \\ 6 & -2 \end{bmatrix}$. The equation $(A - I)\mathbf{x} = \mathbf{0}$ amounts to $6x_1 - 2x_2 = 0$. So $x_1 = (1/3)x_2$, with x_2 free. The general solution is $x_2\begin{bmatrix} 1/3 \\ 1 \end{bmatrix}$. A nice basis for the eigenspace is $\mathbf{u}_1 = \begin{bmatrix} 1 \\ 3 \end{bmatrix}$.

<u>For $\lambda = -1$:</u> $A - (-1)I = \begin{bmatrix} 2 & 0 \\ 6 & 0 \end{bmatrix}$. The equation $(A + I)\mathbf{x} = \mathbf{0}$ amounts to $2x_1 = 0$, with x_2 free. The general solution is $x_2\begin{bmatrix} 0 \\ 1 \end{bmatrix}$. Take $\mathbf{u}_2 = \begin{bmatrix} 0 \\ 1 \end{bmatrix}$ as a basis for the eigenspace.

From $\mathbf{u}_1$ and $\mathbf{u}_2$, construct $P = \begin{bmatrix} \mathbf{u}_1 & \mathbf{u}_2 \end{bmatrix} = \begin{bmatrix} 1 & 0 \\ 3 & 1 \end{bmatrix}$. Then set $D = \begin{bmatrix} 1 & 0 \\ 0 & -1 \end{bmatrix}$, where the eigenvalues in D correspond to $\mathbf{u}_1$ and $\mathbf{u}_2$, respectively.

Warning: The 3×3 matrices in Exercises 12—18 may be diagonalizable even though each matrix has only two distinct eigenvalues. (Theorem 6 gives only a *sufficient* condition for diagonalizability.) You have to check for three linearly independent eigenvectors.

13. $A = \begin{bmatrix} 2 & 2 & -1 \\ 1 & 3 & -1 \\ -1 & -2 & 2 \end{bmatrix}$. The eigenvalues 5 and 1 are given. Because A is 3×3, you need three linearly independent eigenvectors.

For $\lambda = 5$: Solve $(A - 5I)\mathbf{x} = \mathbf{0}$. Form $A - 5I = \begin{bmatrix} -3 & 2 & -1 \\ 1 & -2 & -1 \\ -1 & -2 & -3 \end{bmatrix}$ and compute

$$\begin{bmatrix} -3 & 2 & -1 & 0 \\ 1 & -2 & -1 & 0 \\ -1 & -2 & -3 & 0 \end{bmatrix} \sim \begin{bmatrix} 1 & -2 & -1 & 0 \\ -3 & 2 & -1 & 0 \\ -1 & -2 & -3 & 0 \end{bmatrix} \sim \cdots \sim \begin{bmatrix} 1 & 0 & 1 & 0 \\ 0 & 1 & 1 & 0 \\ 0 & 0 & 0 & 0 \end{bmatrix}$$

So $x_1 = -x_3$, $x_2 = -x_3$, with x_3 free. Take $\mathbf{v}_1 = \begin{bmatrix} -1 \\ -1 \\ 1 \end{bmatrix}$, for instance, as the basis for the eigenspace. At this point, you don't know if A is diagonalizable. Your only hope is to find two linearly independent eigenvectors inside the eigenspace for $\lambda = 1$, because there are no other eigenspaces in which to look.

For $\lambda = 1$: Form $A - I = \begin{bmatrix} 1 & 2 & -1 \\ 1 & 2 & -1 \\ -1 & -2 & 1 \end{bmatrix}$. The equation $(A - I)\mathbf{x} = \mathbf{0}$ reduces to $x_1 + 2x_2 - x_3 = 0$. So $x_1 = -2x_2 + x_3$, with x_2 and x_3 free. At this point you know that the eigenspace is two-dimensional (because there are two free variables). So there are the necessary two linearly independent eigenvectors, and hence A is diagonalizable. To produce the eigenvectors, write the solution of $(A - I)\mathbf{x} = \mathbf{0}$ in the form

$$\begin{bmatrix} x_1 \\ x_2 \\ x_3 \end{bmatrix} = \begin{bmatrix} -2x_2+x_3 \\ x_2 \\ x_3 \end{bmatrix} = x_2\begin{bmatrix} -2 \\ 1 \\ 0 \end{bmatrix} + x_3\begin{bmatrix} 1 \\ 0 \\ 1 \end{bmatrix}$$

Set $\mathbf{v}_2 = \begin{bmatrix} -2 \\ 1 \\ 0 \end{bmatrix}$, $\mathbf{v}_3 = \begin{bmatrix} 1 \\ 0 \\ 1 \end{bmatrix}$, and $P = [\mathbf{v}_1 \quad \mathbf{v}_2 \quad \mathbf{v}_3] = \begin{bmatrix} -1 & -2 & 1 \\ -1 & 1 & 0 \\ 1 & 0 & 1 \end{bmatrix}$.

(The text's answer shows a slightly different but also suitable choice for P.) The columns of P are linearly independent, by Theorem 7, because the eigenvectors form bases for their respective eigenspaces. So P is invertible. Since the first column of P corresponds to $\lambda = 5$, the first diagonal entry in D must be 5, which means that

$$D = \begin{bmatrix} 5 & 0 & 0 \\ 0 & 1 & 0 \\ 0 & 0 & 1 \end{bmatrix}$$

Warning: A common mistake in Exercise 13 is to build a 3×3 matrix P in the usual way and then take D to be a 2×2 matrix, such as

$$D = \begin{bmatrix} 5 & 0 \\ 0 & 1 \end{bmatrix}$$

Of course this doesn't make sense because PD isn't even defined. Another mistake is to make a 2×2 matrix D as above and build P with only two of the three eigenvectors, say one from each eigenspace. Now $AP = PD$, but P is a 3×2 matrix and is not invertible.

19. $A = \begin{bmatrix} 5 & -3 & 0 & 9 \\ 0 & 3 & 1 & -2 \\ 0 & 0 & 2 & 0 \\ 0 & 0 & 0 & 2 \end{bmatrix}$. The eigenvalues are obviously 5, 3, and 2. (Why?)

For $\lambda = 2$: To solve $(A - 2I)\mathbf{x} = \mathbf{0}$, completely reduce $[(A-2I) \;\; \mathbf{0}]$:

$$\begin{bmatrix} 3 & -3 & 0 & 9 & 0 \\ 0 & 1 & 1 & -2 & 0 \\ 0 & 0 & 0 & 0 & 0 \\ 0 & 0 & 0 & 0 & 0 \end{bmatrix} \sim \begin{bmatrix} 3 & 0 & 3 & 3 & 0 \\ 0 & 1 & 1 & -2 & 0 \\ 0 & 0 & 0 & 0 & 0 \\ 0 & 0 & 0 & 0 & 0 \end{bmatrix} \sim \begin{bmatrix} 1 & 0 & 1 & 1 & 0 \\ 0 & 1 & 1 & -2 & 0 \\ 0 & 0 & 0 & 0 & 0 \\ 0 & 0 & 0 & 0 & 0 \end{bmatrix}$$

So $x_1 = -x_3 - x_4$, $x_2 = -x_3 + 2x_4$, with x_3 and x_4 free. The usual calculations produce a basis for the eigenspace:

$$\begin{bmatrix} x_1 \\ x_2 \\ x_3 \\ x_4 \end{bmatrix} = \begin{bmatrix} -x_3 - x_4 \\ -x_3 + 2x_4 \\ x_3 \\ x_4 \end{bmatrix} = x_3\begin{bmatrix} -1 \\ -1 \\ 1 \\ 0 \end{bmatrix} + x_4\begin{bmatrix} -1 \\ 2 \\ 0 \\ 1 \end{bmatrix}. \quad \text{Basis: } \mathbf{v}_1 = \begin{bmatrix} -1 \\ -1 \\ 1 \\ 0 \end{bmatrix}, \; \mathbf{v}_2 = \begin{bmatrix} -1 \\ 2 \\ 0 \\ 1 \end{bmatrix}$$

Checkpoint: If you happened to choose $\lambda = 2$ first, as in this solution, would you have enough information at this point to determine whether A is diagonalizable?

For $\lambda = 3$: To solve $(A - 3I)\mathbf{x} = \mathbf{0}$, completely reduce $[(A-3I) \;\; \mathbf{0}]$:

$$\begin{bmatrix} 2 & -3 & 0 & 9 & 0 \\ 0 & 0 & 1 & -2 & 0 \\ 0 & 0 & -1 & 0 & 0 \\ 0 & 0 & 0 & -1 & 0 \end{bmatrix} \sim \cdots \sim \begin{bmatrix} 1 & -3/2 & 0 & 0 & 0 \\ 0 & 0 & 1 & 0 & 0 \\ 0 & 0 & 0 & 1 & 0 \\ 0 & 0 & 0 & 0 & 0 \end{bmatrix}, \quad \begin{cases} x_1 = (3/2)x_2 \\ x_2 \text{ is free} \\ x_3 = 0 \\ x_4 = 0 \end{cases}$$

Choosing $x_2 = 2$ produces the eigenvector $\mathbf{v}_3 = \begin{bmatrix} 3 \\ 2 \\ 0 \\ 0 \end{bmatrix}$.

For $\lambda = 5$: To solve $(A - 3I)\mathbf{x} = \mathbf{0}$, completely reduce $[(A-5I) \quad \mathbf{0}]$:

$$\begin{bmatrix} 0 & -3 & 0 & 9 & 0 \\ 0 & -2 & 1 & -2 & 0 \\ 0 & 0 & -3 & 0 & 0 \\ 0 & 0 & 0 & -3 & 0 \end{bmatrix} \sim \cdots \sim \begin{bmatrix} 0 & 1 & 0 & 0 & 0 \\ 0 & 0 & 1 & 0 & 0 \\ 0 & 0 & 0 & 1 & 0 \\ 0 & 0 & 0 & 0 & 0 \end{bmatrix}, \quad \begin{cases} x_1 \text{ is free} \\ x_2 = 0 \\ x_3 = 0 \\ x_4 = 0 \end{cases}$$

A basis for the eigenspace is $\mathbf{v}_4 = \begin{bmatrix} 1 \\ 0 \\ 0 \\ 0 \end{bmatrix}$. From the work above, set

$$P = [\mathbf{v}_1 \quad \mathbf{v}_2 \quad \mathbf{v}_3 \quad \mathbf{v}_4] = \begin{bmatrix} -1 & -1 & 3 & 1 \\ -1 & 2 & 2 & 0 \\ 1 & 0 & 0 & 0 \\ 0 & 1 & 0 & 0 \end{bmatrix}, \quad D = \begin{bmatrix} 2 & 0 & 0 & 0 \\ 0 & 2 & 0 & 0 \\ 0 & 0 & 3 & 0 \\ 0 & 0 & 0 & 5 \end{bmatrix}$$

This answer differs from that in the text. There, $P = [\mathbf{v}_4 \quad \mathbf{v}_3 \quad \mathbf{v}_1 \quad \mathbf{v}_2]$, and the entries in D are rearranged to match the new order of the eigenvectors. According to the Diagonalization Theorem, both answers are correct.

21. A is diagonalizable because you know that five linearly independent eigenvectors exist: three in the three-dimensional eigenspace and two in the two-dimensional eigenspace. Theorem 7 guarantees that the set of all five eigenvectors is linearly independent.

23. Let $\mathbf{v}_1$ be a basis for the one-dimensional eigenspace, let $\mathbf{v}_2$ and $\mathbf{v}_3$ be a basis for the two-dimensional eigenspace, and let $\mathbf{v}_4$ be any eigenvector in the remaining eigenspace. By Theorem 7, $\{\mathbf{v}_1, \mathbf{v}_2, \mathbf{v}_3, \mathbf{v}_4\}$ is linearly independent. Since A is 4×4, the Diagonalization Theorem shows that A is diagonalizable.

25. If A is diagonalizable, then $A = PDP^{-1}$ for some invertible P and diagonal D. Since A is invertible, 0 is not an eigenvalue of A. So the diagonal entries in D (which are eigenvalues of A) are not zero, and D is invertible. By the theorem on the inverse of a product,

$$A^{-1} = (PDP^{-1})^{-1} = (P^{-1})^{-1}D^{-1}P^{-1} = PD^{-1}P^{-1}$$

Since D^{-1} is obviously diagonal, A^{-1} is diagonalizable.

27. The diagonal entries in D_1 are reversed from those in D. So interchange the (eigenvector) columns of P to make them correspond properly to the eigenvalues in D_1. In this case,

$$P_1 = \begin{bmatrix} 1 & 1 \\ -2 & -1 \end{bmatrix} \quad \text{and} \quad D_1 = \begin{bmatrix} 3 & 0 \\ 0 & 5 \end{bmatrix}$$

Although the first column of P must be an eigenvector corresponding to the eigenvalue 3, there is nothing to prevent us from selecting some multiple of $\begin{bmatrix} 1 \\ -2 \end{bmatrix}$, say $\begin{bmatrix} -3 \\ 6 \end{bmatrix}$, and letting $P_2 = \begin{bmatrix} -3 & 1 \\ 6 & -1 \end{bmatrix}$. We now have three different factorizations or "diagonalizations" of A:

$$A = PDP^{-1} = P_1D_1P_1^{-1} = P_2D_1P_2^{-1}$$

Answer to Checkpoint: Yes. In Exercise 19, the fact that the eigenspace for $\lambda = 2$ is two-dimensional guarantees that A is diagonalizable, because each of the other two eigenvalues will produce at least one eigenvector, and the resulting set of four eigenvectors will be linearly independent, by Theorem 7. So A is diagonalizable, by the Diagonalization Theorem. (Note that since one eigenspace is two-dimensional, the other eigenspaces must be one-dimensional, because there could not possibly be *more* than four linearly independent eigenvectors in $\mathbb{R}^4$.)

MATLAB The command **eig(A)** produces a vector containing the eigenvalues of A. More generally, the command **[P D] = eig(A)** produces two square matrices P and D (or any other names you choose) such that $AP = PD$, with D diagonal. If A happens to be diagonalizable, then P will be invertible. In any case, P is likely to be quite different from what you construct for the homework. The columns of P are scaled so they are "unit" vectors (such that the sum of squares of the entries is 1). Dividing P by one of its entries may make some of the new entries recognizable, because the problems in the text usually involve simple numbers.

6.4 EIGENVALUES AND LINEAR TRANSFORMATIONS

This section introduces the matrix of a linear transformation relative to specified bases for vector spaces, and then uses this concept to give a new interpretation of the matrix factorization $A = PDP^{-1}$.

STUDY NOTES

The exercises will help you learn the definition of the matrix representation of a transformation relative to specified bases, say $\mathcal{B}$ and $\mathcal{C}$. Compare this definition with the standard matrix of a transformation from $\mathbb{R}^n$ into $\mathbb{R}^m$. What are $\mathcal{B}$ and $\mathcal{C}$ in this case?

After Example 1 (and Exercises 1-6 and 9-10), the section focuses on the case when $T: V \to V$ and $\mathcal{B} = \mathcal{C}$. The following algorithm and the solution to Exercise 7 describe two different ways to construct $[T]_{\mathcal{B}}$.

Algorithm for Finding the B-matrix of T:V → V

1. Compute the images of the basis vectors:

$$T(\mathbf{b}_1), \ \ldots, \ T(\mathbf{b}_n)$$

2. Convert these images into $\mathcal{B}$-coordinate vectors:

$$[T(\mathbf{b}_1)]_{\mathcal{B}}, \ \ldots, \ [T(\mathbf{b}_n)]_{\mathcal{B}}$$

3. Place the $\mathcal{B}$-coordinate vectors into the columns of $[T]_{\mathcal{B}}$.

The sentence before Theorem 8 summarizes the main idea of the theorem. Studying the proof should help you understand the theorem and review important concepts. The subsection on Similarity of Matrix Representations could have contained more ideas. Take notes carefully if your instructor decides to expand this material somewhat.

SOLUTIONS TO EXERCISES

1. $T(\mathbf{b}_1) = 3\mathbf{d}_1 - 5\mathbf{d}_2$, $T(\mathbf{b}_2) = -\mathbf{d}_1 + 6\mathbf{d}_2$, $T(\mathbf{b}_3) = 4\mathbf{d}_2$

$$[T(\mathbf{b}_1)]_{\mathcal{D}} = \begin{bmatrix} 3 \\ -5 \end{bmatrix} \qquad [T(\mathbf{b}_2)]_{\mathcal{D}} = \begin{bmatrix} -1 \\ 6 \end{bmatrix} \qquad [T(\mathbf{b}_3)]_{\mathcal{D}} = \begin{bmatrix} 0 \\ 4 \end{bmatrix}$$

Matrix for T relative to $\mathcal{B}$ and $\mathcal{D}$: $\begin{bmatrix} 3 & -1 & 0 \\ -5 & 6 & 4 \end{bmatrix}$

7. The method of Example 2 works. But, perhaps a faster way is to realize that the information given provides the general form of $T(\mathbf{p})$ as shown in the figure below:

$$a_0 + a_1 t + a_2 t^2 \xrightarrow{\quad T \quad} 3a_0 + (5a_0 - 2a_1)t + (4a_1 + a_2)t^2$$

coordinate mapping $\downarrow$ $\qquad$ coordinate mapping $\downarrow$

$$\begin{bmatrix} a_0 \\ a_1 \\ a_2 \end{bmatrix} \xrightarrow[\text{by } [T]_B]{\text{multiplication}} \begin{bmatrix} 3a_0 \\ 5a_0 - 2a_1 \\ 4a_1 + a_2 \end{bmatrix}$$

The matrix that implements the multiplication along the bottom of the figure is easily filled in by inspection:

$$\begin{bmatrix} ? & ? & ? \\ ? & ? & ? \\ ? & ? & ? \end{bmatrix}\begin{bmatrix} a_0 \\ a_1 \\ a_2 \end{bmatrix} = \begin{bmatrix} 3a_0 \\ 5a_0 - 2a_1 \\ 4a_1 + a_2 \end{bmatrix} \quad \text{implies that } [T]_{\mathcal{B}} = \begin{bmatrix} 3 & 0 & 0 \\ 5 & -2 & 0 \\ 0 & 4 & 1 \end{bmatrix}$$

Study Tip: See the Study Guide notes for Section 2.6 (particularly for Exercise 19). This method allows you to find a matrix that implements the mapping without assuming that T is linear. In fact, this derivation *proves* that T is linear, because it is now represented as a matrix transformation. Why not try this method on Example 2?

13. Start by diagonalizing $A = \begin{bmatrix} 0 & 1 \\ -3 & 4 \end{bmatrix}$. The characteristic polynomial is $\lambda^2 - 4\lambda + 3 = (\lambda - 1)(\lambda - 3)$, so the eigenvalues are 1 and 3.

<u>For $\lambda = 1$:</u> $A - I = \begin{bmatrix} -1 & 1 \\ -3 & 3 \end{bmatrix}$. The equation $(A - I)\mathbf{x} = \mathbf{0}$ amounts to $-x_1 + x_2 = 0$. So $x_1 = x_2$, with x_2 free. As a basis, take $\mathbf{u}_1 = \begin{bmatrix} 1 \\ 1 \end{bmatrix}$.

<u>For $\lambda = 3$:</u> $A - 3I = \begin{bmatrix} -3 & 1 \\ -3 & 1 \end{bmatrix}$. The equation $(A - I)\mathbf{x} = \mathbf{0}$ amounts to $-3x_1 + x_2 = 0$. So $x_1 = x_2/3$, with x_2 free. As a basis, take $\mathbf{u}_2 = \begin{bmatrix} 1 \\ 3 \end{bmatrix}$.

The vectors $\mathbf{u}_1, \mathbf{u}_2$ can form the columns of a matrix P that diagonalizes A. By Theorem 8, the basis $\mathcal{B} = \{\mathbf{u}_1, \mathbf{u}_2\}$ has the property that the $\mathcal{B}$-matrix of the transformation $\mathbf{x} \mapsto A\mathbf{x}$ is a diagonal matrix.

19. If A is similar to B, then there exists an invertible matrix P such that $P^{-1}AP = B$. Thus B is invertible because it is the product of invertible matrices. By a theorem about inverses of products, $B^{-1} = P^{-1}A^{-1}(P^{-1})^{-1} = P^{-1}A^{-1}P$, which shows that A^{-1} is similar to B^{-1}.

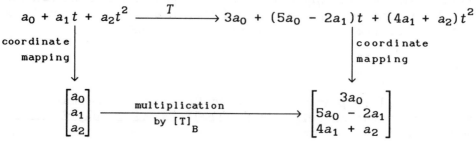

21. By hypothesis, there exist invertible P and Q such that $P^{-1}BP = A$ and $Q^{-1}CQ = A$. Then $P^{-1}BP = Q^{-1}CQ$. Left-multiplying by Q, and right-multiplying by Q^{-1}, we have $QP^{-1}BPQ^{-1} = QQ^{-1}CQQ^{-1}$. So $C = QP^{-1}BPQ^{-1} = (PQ^{-1})^{-1}B(PQ^{-1})$, which shows that B is similar to C.

23. If $A\mathbf{x} = \lambda\mathbf{x}$, $\mathbf{x} \neq \mathbf{0}$, then $P^{-1}A\mathbf{x} = \lambda P^{-1}\mathbf{x}$. If $B = P^{-1}AP$, then

$$B(P^{-1}\mathbf{x}) = P^{-1}AP(P^{-1}\mathbf{x}) = P^{-1}A\mathbf{x} = \lambda P^{-1}\mathbf{x} \qquad (*)$$

by the first calculation. Note that $P^{-1}\mathbf{x} \neq \mathbf{0}$, because $\mathbf{x} \neq \mathbf{0}$ and P^{-1} is invertible. Hence (*) shows that $P^{-1}\mathbf{x}$ is an eigenvector of B corresponding to λ. (Of course, λ is an eigenvalue of both A and B because the matrices are similar, by Theorem 4 in Section 6.2.)

25. If $A = PBP^{-1}$, then

$$\text{tr}(A) = \text{tr}((PB)P^{-1}) = \text{tr}(P^{-1}(PB)) \quad \text{By the trace property}$$

$$= \text{tr}(P^{-1}PB) = \text{tr}(IB) = \text{tr}(B)$$

If B is diagonal, then the diagonal entries of B must be the eigenvalues of A, by the diagonalization theorem (Theorem 5 in Section 6.3). So $\text{tr}\,A = \text{tr}\,B = \{\text{sum of the eigenvalues of } A\}$.

Study Tip: The result of Exercise 25 holds for any square matrix A. That is, the trace of A is the sum of the eigenvalues of A, counted according to multiplicities. Use this fact to provide a quick check on your eigenvalue calculations. The sum of the eigenvalues must match the sum of the diagonal entries in A (even if A is not diagonalizable).

29. If $\mathcal{B} = \{\mathbf{b}_1, \ldots, \mathbf{b}_n\}$, then the $\mathcal{B}$-coordinate vector of $\mathbf{b}_j$ is $\mathbf{e}_j$, the standard basis vector for $\mathbb{R}^n$. For instance,

$$\mathbf{b}_1 = 1\cdot\mathbf{b}_1 + 0\cdot\mathbf{b}_2 + \cdots + 0\cdot\mathbf{b}_n$$

Thus $[I(\mathbf{b}_j)]_{\mathcal{B}} = [\mathbf{b}_j]_{\mathcal{B}} = \mathbf{e}_j$, and

$$[I]_{\mathcal{B}} = [[I(\mathbf{b}_1)]_{\mathcal{B}} \quad \cdots \quad [I(\mathbf{b}_n)]_{\mathcal{B}}] = [\mathbf{e}_1 \quad \cdots \quad \mathbf{e}_n] = I$$

Appendix: The Characteristic Polynomial of a 2×2 Matrix

By Exercise 19 in Section 6.2, $\det A$ is the product of the eigenvalues of A. From the Study Tip above, $\text{tr}\,A$ is the sum of the eigenvalues. Thus, if a 2×2 matrix A has eigenvalues λ_1 and λ_2, the characteristic polynomial of A has the form

$$(\lambda_1 - \lambda)(\lambda_2 - \lambda) = \lambda^2 - (\lambda_1 + \lambda_2)\lambda + \lambda_1\lambda_2$$
$$= \lambda^2 - (\operatorname{tr} A)\lambda + (\det A)$$

6.5 COMPLEX EIGENVALUES

If the characteristic equation of an $n \times n$ real matrix A has a complex eigenvalue λ and if $\mathbf{v}$ is a nonzero vector in $\mathbb{C}^n$ such that $A\mathbf{v} = \lambda\mathbf{v}$, then both λ and $\mathbf{v}$ provide useful information about A.

KEY IDEAS

Only matrices with real entries are considered here. If λ is a complex eigenvalue of A, with $\mathbf{v}$ a corresponding eigenvector, then $\bar{\lambda}$ is also an eigenvalue of A, with $\bar{\mathbf{v}}$ an eigenvector. Find out if you should know how to prove this fact.

Example 6 describes the prototype for all 2×2 matrices with a complex eigenvalue λ. Only λ is needed if you only want to know the angle φ of rotation and the scale factor $|\lambda|$, but an associated eigenvector $\mathbf{v}$ is also needed if you want to factor A as PCP^{-1}, as in Example 7.

SOLUTIONS TO EXERCISES

1. $A = \begin{bmatrix} 1 & -2 \\ 1 & 3 \end{bmatrix}$, $A - \lambda I = \begin{bmatrix} 1-\lambda & -2 \\ 1 & 3-\lambda \end{bmatrix}$

$$\det(A - \lambda I) = (1 - \lambda)(3 - \lambda) - (-2) = \lambda^2 - 4\lambda + 5$$

Use the quadratic formula to find the eigenvalues: $\lambda = \dfrac{4 \pm \sqrt{16 - 20}}{2} =$ $2 \pm i$. Example 2 gives a shortcut for finding one eigenvector, and Example 5 shows how to write the other eigenvector with no effort.

For $\lambda = 2 + i$: $A - (2 + i)I = \begin{bmatrix} -1-i & -2 \\ 1 & 1-i \end{bmatrix}$. The equation $(A - \lambda I)\mathbf{x} = \mathbf{0}$ gives

$$\begin{aligned} (-1 - i)x_1 - \quad\quad 2x_2 &= 0 \\ x_1 + (1 - i)x_2 &= 0 \end{aligned}$$

As in Example 2, the two equations are equivalent—each determines the same relation between x_1 and x_2. So use the second equation to obtain $x_1 = -(1 - i)x_2$, with x_2 free. The general solution is $x_2\begin{bmatrix} -1 + i \\ 1 \end{bmatrix}$, and the vector $\mathbf{v}_1 = \begin{bmatrix} -1 + i \\ 1 \end{bmatrix}$ is a basis for the eigenspace.

<u>For $\lambda = 2 - i$:</u> Let $\mathbf{v}_2 = \overline{\mathbf{v}_1} = \begin{bmatrix} -1 - i \\ 1 \end{bmatrix}$. The remark prior to Example 5 shows that $\mathbf{v}_2$ is automatically an eigenvector for $\overline{2 + i}$. In fact, calculations similar to those above would show that $\mathbf{v}_2$ is a basis for the eigenspace. (In general, for a real matrix A, it can be shown that forming the complex conjugates of a basis of the eigenspace for λ produces a basis of the eigenspace for $\bar{\lambda}$.)

7. $A = \begin{bmatrix} \sqrt{3} & -1 \\ 1 & \sqrt{3} \end{bmatrix}$. The eigenvalues are $\sqrt{3} \pm i$. Ask your instructor if you are permitted to write down the eigenvalues of $\begin{bmatrix} a & -b \\ b & a \end{bmatrix}$ from memory, or if you are expected to find them via the characteristic equation. Note that the eigenvectors are easy to remember, too. See the Practice Problem.

The scale factor associated with the transformation $\mathbf{x} \mapsto A\mathbf{x}$ is simply $r = |\lambda| = ((\sqrt{3})^2 + 1^2)^{1/2} = 2$. For the angle of the rotation, plot the point $(a,b) = (\sqrt{3},1)$ in the xy-plane and use trigonometry.

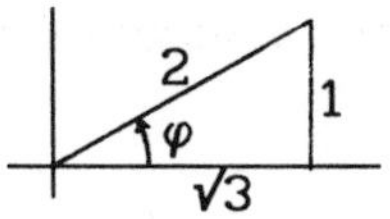

$\varphi = \pi/6$ radians

13. From Exercise 1, $\lambda = 2 \pm i$, and the eigenvector $\mathbf{v} = \begin{bmatrix} -1 - i \\ 1 \end{bmatrix}$ corresponds to $\lambda = 2 - i$. Since $\text{Re}\,\mathbf{v} = \begin{bmatrix} -1 \\ 1 \end{bmatrix}$ and $\text{Im}\,\mathbf{v} = \begin{bmatrix} -1 \\ 0 \end{bmatrix}$, take $P = \begin{bmatrix} -1 & -1 \\ 1 & 0 \end{bmatrix}$. Then compute

$$C = P^{-1}AP = \begin{bmatrix} 0 & 1 \\ -1 & -1 \end{bmatrix}\begin{bmatrix} 1 & -2 \\ 1 & 3 \end{bmatrix}\begin{bmatrix} -1 & -1 \\ 1 & 0 \end{bmatrix} = \begin{bmatrix} 0 & 1 \\ -1 & -1 \end{bmatrix}\begin{bmatrix} -3 & -1 \\ 2 & -1 \end{bmatrix} = \begin{bmatrix} 2 & -1 \\ 1 & 2 \end{bmatrix}$$

Actually, Theorem 9 gives the formula for C. Note that the eigenvector $\mathbf{v}$ corresponds to $a - bi$ instead of $a + bi$. If, for instance, you use the eigenvector for $2 + i$, your C will be $\begin{bmatrix} 2 & 1 \\ -1 & 2 \end{bmatrix}$. The imaginary part of the eigenvalue is the (1,2)-entry in C.

So there are two possible choices for C (depending on the vector used to produce P). On an exam, if you are not sure of the form of C, you can always compute it quickly from the formula $C = P^{-1}AP$, as in the solution above.

Note: Because there are two possibilities for C in the factorization of a 2×2 matrix as in Exercise 13, the measure of rotation φ associated with the transformation $\mathbf{x} \mapsto A\mathbf{x}$ is determined only up to a change of sign. The "orientation" of the angle is determined by the change of variable $\mathbf{x} = P\mathbf{u}$. See Fig. 4 in the text.

19. $A = \begin{bmatrix} 1.52 & -.7 \\ .56 & .4 \end{bmatrix}$, $\det(A - \lambda I) = \lambda^2 - 1.92\lambda + 1$.

Use the quadratic formula to solve $\lambda^2 - 1.92\lambda + 1 = 0$:

$$\lambda = \frac{1.92 \pm \sqrt{-.3136}}{2} = .96 \pm .28i$$

To find the eigenvector for $\lambda = .96 - .28i$, solve

$$\begin{aligned} (1.52 - .96 + .28i)x_1 \qquad\qquad & -.7x_2 = 0 \\ .56x_1 + (.4 - .96 + .28i)x_2 &= 0 \end{aligned}$$

There is a nonzero solution (because $.96 - .28i$ is an eigenvalue), so you can use either equation to find the solution. From the second equation,

$$x_1 = \frac{.56 - .28i}{.56}x_2 = (1 - .5i)x_2$$

Setting $x_2 = 2$ produces the (complex) eigenvector $\mathbf{v} = \begin{bmatrix} 2 - i \\ 2 \end{bmatrix}$.

Since $\text{Re}\,\mathbf{v} = \begin{bmatrix} 2 \\ 2 \end{bmatrix}$ and $\text{Im}\,\mathbf{v} = \begin{bmatrix} -1 \\ 0 \end{bmatrix}$, take $P = \begin{bmatrix} 2 & -1 \\ 2 & 0 \end{bmatrix}$. Finally, compute

$$P^{-1}AP = \frac{1}{2}\begin{bmatrix} 0 & 1 \\ -2 & 2 \end{bmatrix}\begin{bmatrix} 1.52 & -.7 \\ .56 & .4 \end{bmatrix}\begin{bmatrix} 2 & -1 \\ 2 & 0 \end{bmatrix} = \begin{bmatrix} .96 & -.28 \\ .28 & .96 \end{bmatrix}$$

This final matrix, which has the proper form, is C.

25. $\text{Re}(A\mathbf{x}) = \frac{1}{2}(A\mathbf{x} + \overline{A\mathbf{x}}) = A[\frac{1}{2}(\mathbf{x} + \overline{\mathbf{x}})]$ Because A is real matrix

$= A(\text{Re}\,\mathbf{x})$

$\text{Im}(A\mathbf{x}) = \frac{1}{2i}(A\mathbf{x} - \overline{A\mathbf{x}}) = A[\frac{1}{2i}(\mathbf{x} - \overline{\mathbf{x}})] = A(\text{Im}\,\mathbf{x})$

MATLAB: The command **[V D] = eig(A)** (mentioned in Section 6.3) works for matrices with complex eigenvalues. In this case V and the diagonal matrix D have some complex entries. For a 2×2 real matrix with a complex eigenvalue, MATLAB tends to place the eigenvalue $a - bi$ (where $b > 0$) as the (2,2)-entry of D. MATLAB does not produce matrices for a factorization $A = PCP^{-1}$ of the sort described in this section.

For any matrix V, the commands **real(V)** and **imag(V)** produce the real and imaginary parts of the entries in V, displayed as matrices the same size as V.

6.6 APPLICATIONS TO DYNAMICAL SYSTEMS

This section presents the climax to a crescendo of ideas that began in Section 2.7 and flowed through parts of Chapters 5 and 6. You need not have read all the material to appreciate the interesting applications in this section, but you will profit from a review of page 278 and Example 5 in Section 6.2.

KEY IDEAS

A **solution** of a first order homogeneous difference equation

$$\mathbf{x}_{k+1} = A\mathbf{x}_k \quad (k = 0, 1, 2, \ldots) \tag{1}$$

is a sequence $\{\mathbf{x}_k\}$ that satisfies (1) and is described by a formula for each $\mathbf{x}_k$ that does not depend on the preceding terms in the sequence other than the initial term $\mathbf{x}_0$. In Section 6.1, you saw how a solution can be constructed when $\mathbf{x}_0$ is an eigenvector. When $\mathbf{x}_0$ is *not* an eigenvector, look for an eigenvector decomposition of $\mathbf{x}_0$:

$$\mathbf{x}_0 = c_1\mathbf{v}_1 + \cdots + c_n\mathbf{v}_n \qquad \text{Each } \mathbf{v}_i \text{ is an eigenvector.} \tag{2}$$

To make (2) possible for any $\mathbf{x}_0$ in $\mathbb{R}^n$, the section assumes that the $n \times n$ matrix A has n linearly independent eigenvectors. If $\mathbf{x}_k = A^k\mathbf{x}_0$, then

$$\mathbf{x}_k = c_1(\lambda_1)^k\mathbf{v}_1 + \cdots + c_n(\lambda_n)^k\mathbf{v}_n$$

When $\{\mathbf{x}_k\}$ describes the "state" of a system at discrete times (denoted by $k = 0,1,2,\ldots$), the *long-term behavior* of this dynamical system is a description of what happens to $\mathbf{x}_k$ as $k \to \infty$. The text focuses on the following important situation.

Let A be an $n \times n$ matrix with n linearly independent eigenvectors, corresponding to eigenvalues such that $|\lambda_1| \geq 1 > |\lambda_j|$ for $j = 2,\dots,n$. If $\mathbf{x}_0$ is given by (2) with $c_1 \neq 0$, then for all sufficiently large k,

$\mathbf{x}_{k+1} \approx \lambda_1 \mathbf{x}_k$	Each entry in $\mathbf{x}_k$ grows by a factor of λ_1.
$\mathbf{x}_k \approx c_1(\lambda_1)^k \mathbf{v}_1$	$\mathbf{x}_k$ is approximately a multiple of $\mathbf{v}_1$, and so the ratio between any two entries in $\mathbf{x}_k$ is nearly the same as the corresponding ratio for $\mathbf{v}_1$.

STUDY NOTES

When the two approximations above are true in an application, the eigenvalue λ_1 and the eigenvector $\mathbf{v}_1$ have interesting physical interpretations. Make sure you can describe this on an exam. (See the last three sentences in the solution of Example 1, for instance.)

The predator-prey model is rather primitive and provides only a starting point for more refined models. Still, you might enjoy considering what the model in Example 1 predicts if $\mathbf{x}_0$ happens to be a multiple of $\mathbf{v}_2 = (5, 1)$, or if initially there are *more* than 5 owls for every 1 thousand rats, assuming $p = .104$. Incidentally, the predation parameter p is the average number of rats (in thousands) eaten by one owl in one month.

The graphical descriptions of solutions to difference equations should help you understand what can happen to $\mathbf{x}_k$ as $k \to \infty$. I hope you enjoy studying the figures even if your class does not have time to cover this part of the section. Only the simplest cases are shown, but these cases form the foundation for studying *nonlinear* dynamical systems which are widely used (but require calculus techniques not covered here). Even for nonlinear systems, eigenvalues and eigenvectors of certain matrices play an important role.

SOLUTIONS TO EXERCISES

1. a. The eigenvectors $\mathbf{v}_1 = \begin{bmatrix}1\\1\end{bmatrix}$ and $\mathbf{v}_2 = \begin{bmatrix}-1\\1\end{bmatrix}$ form a basis for $\mathbb{R}^2$. To find the action of A on $\mathbf{x}_0 = \begin{bmatrix}9\\1\end{bmatrix}$, express $\mathbf{x}_0$ in terms of $\mathbf{v}_1$ and $\mathbf{v}_2$. That is, find c_1, c_2 such that $\mathbf{x}_0 = c_1\mathbf{v}_1 + c_2\mathbf{v}_2$:

$$\begin{bmatrix}1 & -1 & 9\\1 & 1 & 1\end{bmatrix} \sim \begin{bmatrix}1 & 0 & 5\\0 & 1 & -4\end{bmatrix} \Rightarrow \mathbf{x}_0 = 5\mathbf{v}_1 - 4\mathbf{v}_2$$

Since $\mathbf{v}_1$, $\mathbf{v}_2$ are eigenvectors (for the eigenvalues 3 and 1/3):

$$\mathbf{x}_1 = A\mathbf{x}_0 = 5A\mathbf{v}_1 - 4A\mathbf{v}_2 = 5\cdot3\mathbf{v}_1 - 4\cdot(1/3)\mathbf{v}_2$$

$$= \begin{bmatrix}15\\15\end{bmatrix} - \begin{bmatrix}-4/3\\4/3\end{bmatrix} = \begin{bmatrix}49/3\\41/3\end{bmatrix}$$

b. Each time A acts on a linear combination of $\mathbf{v}_1$ and $\mathbf{v}_2$, the $\mathbf{v}_1$ term is multiplied by the eigenvalue 3 and the $\mathbf{v}_2$ term is multiplied by the eigenvalue 1/3.

$$\mathbf{x}_2 = A\mathbf{x}_1 = A[5(3)\mathbf{v}_1 - 4(1/3)\mathbf{v}_2] = 5(3)^2\mathbf{v}_1 - 4(1/3)^2\mathbf{v}_2$$

In general, $\mathbf{x}_k = 5(3)^k\mathbf{v}_1 - 4(1/3)^k\mathbf{v}_2$ for $k \geq 0$.

7. a. The matrix A in Exercise 1 has eigenvalues 3 and 1/3. Since $|3| > 1$ and $|1/3| < 1$, the origin is a saddle point.

b. The direction of greatest attraction is determined by $\mathbf{v}_2 = \begin{bmatrix}-1\\1\end{bmatrix}$, the eigenvector corresponding to the smaller eigenvalue. The direction of greatest repulsion is determined by $\mathbf{v}_1 = \begin{bmatrix}1\\1\end{bmatrix}$, the eigenvector corresponding to the eigenvalue greater than 1.

c. The drawing below shows: (1) lines through the eigenvectors and the origin, (2) arrows toward the origin (showing attraction) on the line through $\mathbf{v}_2$ and arrows away from the origin (showing repulsion) on the line through $\mathbf{v}_1$, (3) several typical trajectories (with arrows) that show the general flow of points. No specific points other than $\mathbf{v}_1$ and $\mathbf{v}_2$ were computed. This type of drawing is about all that one can make without using a computer to plot points.

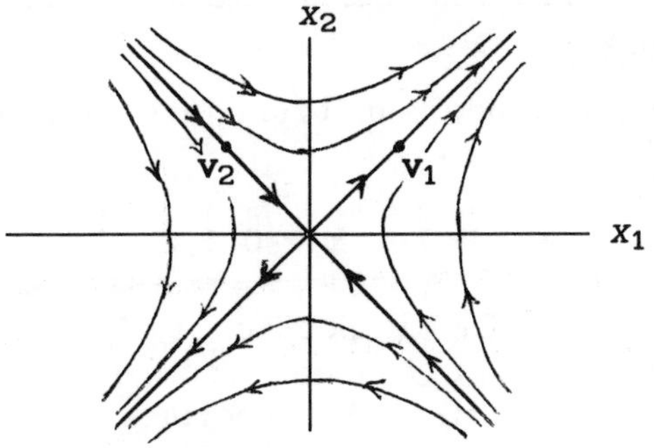

Remark: Sketching trajectories for a dynamical system in which the origin is an attractor or a repellor is more difficult than the sketch in Exercise 7. There has been no discussion of the direction in which the trajectories "bend" as they move toward or away from the origin. For instance, if you rotate Figure 1 of Section 6.6 through a quarter-turn and relabel the axes so that x_1 is on the horizontal axis, then the new figure corresponds to the matrix A with the diagonal entries .8 and .64 interchanged. In general, if A is a diagonal matrix, with positive diagonal entries a and d, unequal to 1, then the trajectories lie on the axes or on curves whose equations have the form $x_2 = r(x_1)^s$, where $s = (\ln d)/(\ln a)$ and r depends on the initial point $\mathbf{x}_0$. (See *Encounters with Chaos*, by Denny Gulick, New York: McGraw-Hill, 1992, pp. 147-150.)

Study Tip: If your instructor wants you to graph trajectories when the origin is an attractor or repellor, there will need to be some class discussion of exactly what you are expected to do.

13. $A = \begin{bmatrix} .8 & .3 \\ -.4 & 1.5 \end{bmatrix}$. First find the eigenvalues:

$$\begin{aligned} \det(A - \lambda I) &= (.8 - \lambda)(1.5 - \lambda) - (.3)(-.4) = \lambda^2 - 2.3 + 1.32 \\ &= (\lambda - 1.2)(\lambda - 1.1) \quad \text{Use the quadratic formula, if needed.} \\ &= 0 \end{aligned}$$

Since both eigenvalues, 1.2 and 1.1, are greater than 1, the origin is a repellor. For the direction of greatest repulsion, find the eigenvector for the larger eigenvalue, 1.2:

$$[(A - 1.2I) \quad \mathbf{0}] = \begin{bmatrix} -.4 & .3 & 0 \\ -.4 & .3 & 0 \end{bmatrix} \sim \begin{bmatrix} 1 & -3/4 & 0 \\ 0 & 0 & 0 \end{bmatrix}, \quad \mathbf{x} = x_2\begin{bmatrix} 3/4 \\ 1 \end{bmatrix}$$

Any multiple of $\begin{bmatrix} 3/4 \\ 1 \end{bmatrix}$, such as $\begin{bmatrix} 3 \\ 4 \end{bmatrix}$, determines the direction of greatest repulsion.

Appendix: The Nonhomogeneous Case

Some applications of difference equations are modeled by a nonhomogeneous equation of the form

$$\mathbf{x}_{k+1} = A\mathbf{x}_k + \mathbf{b}, \quad k = 0, 1, 2, \ldots$$

For instance, see equation (6) in Section 2.7. Assume that A is a 2×2 matrix with eigenvalues λ_1, λ_2 different from 1 and such that $|\lambda_1| > |\lambda_2|$, and let $\mathbf{v}_1$ and $\mathbf{v}_2$ be corresponding eigenvectors.

To solve the difference equation, first find a vector $\mathbf{p}$ such that $\mathbf{p} = A\mathbf{p} + \mathbf{b}$, which is equivalent to solving

$$(I - A)\mathbf{p} = \mathbf{b} \tag{3}$$

Since 1 is not an eigenvalue of A, the matrix $I - A$ is invertible, and (3) has a solution. Next, given any initial point $\mathbf{x}_0$, find c_1, c_2 such that

$$\mathbf{x}_0 = c_1\mathbf{v}_1 + c_2\mathbf{v}_2 + \mathbf{p}$$

(That is, solve $\mathbf{x}_0 - \mathbf{p} = c_1\mathbf{v}_1 + c_2\mathbf{v}_2$ for c_1, c_2.) Finally, compute

$$\begin{aligned}\mathbf{x}_1 &= A\mathbf{x}_0 + \mathbf{b} = A(c_1\mathbf{v}_1 + c_2\mathbf{v}_2 + \mathbf{p}) + \mathbf{b}\\ &= c_1A\mathbf{v}_1 + c_2A\mathbf{v}_2 + A\mathbf{p} + \mathbf{b}\\ &= c_1\lambda_1\mathbf{v}_1 + c_2\lambda_2\mathbf{v}_2 + \mathbf{p}\end{aligned}$$

by definition of $\mathbf{p}$. In general,

$$\mathbf{x}_k = c_1(\lambda_1)^k\mathbf{v}_1 + c_2(\lambda_2)^k\mathbf{v}_2 + \mathbf{p}$$

The sequence $\{\mathbf{x}_k\}$ here is just a translate (by $\mathbf{p}$) of the sequences studied earlier in the section. For instance, if $|\lambda_1| < 1$ and $|\lambda_2| < 1$, then $\mathbf{x}_k$ tends to $\mathbf{p}$ as $k \to \infty$.

MATLAB Given a vector **x**, the command **x = A*x** will compute the "next" point on the trajectory. Use the up-arrow (↑) and <Enter> to repeat the command, over and over.

The following steps create a "trajectory" matrix T whose columns are the points $\mathbf{x}$, $A\mathbf{x}$, $A^2\mathbf{x}, \ldots, A^{15}\mathbf{x}$. (Change 15 to any integer you wish.)

```
T = x              Put x in the first column of T.
for j=1:15         This loop repeats the next two lines 15 times.
  x = A*x;         Compute the next point on the trajectory.
  T = [T  x]       Store the new point in T.
end                End of the loop.
```

After you type the line beginning with "for", MATLAB will suspend all calculations (while you type additional lines) until you type "end".

If you want MATLAB itself to plot the points in T, use the commands:

```
plot(T(1,:),T(2,:),'ow'), grid
```

The 'o' produces a small circle at each point on the trajectory. The 'w' makes the circle white. If you have the data for another trajectory stored in a matrix S, you can plot both trajectories on the same graph:

```
plot(T(1,:),T(2,:),'ow',S(1,:),S(2,:),'*g'), grid    g is for green.
```

Each new "plot" command erases the previous graph.

6.7 ITERATIVE ESTIMATES FOR EIGENVALUES

The algorithms in this section illustrate another use of the eigenvector decomposition described in Section 6.6 (on page 311). Other methods for eigenvalue estimation were mentioned in Section 6.2 (on page 286).

KEY IDEAS

Throughout the section, we suppose that the initial vector $\mathbf{x}_0$ can be written as $\mathbf{x}_0 = c_1\mathbf{v}_1 + \cdots + c_n\mathbf{v}_n$, where $\mathbf{v}_1,\ldots,\mathbf{v}_n$ are eigenvectors of A and $c_1 \neq 0$. (In practice, you will not know $c_1\mathbf{v}_1,\ldots,c_n\mathbf{v}_n$. This eigenvector decomposition is used only to explain why the power method works.)

The Power Method: Assume the eigenvalue λ_1 for $\mathbf{v}_1$ is a strictly dominant eigenvalue (so that $|\lambda_1| > |\lambda_j|$ for $j = 2,\ldots,n$). Then, for large k, the line through $A^k\mathbf{x}_0$ and $\mathbf{0}$ nearly coincides with the line through $\mathbf{v}_1$ and $\mathbf{0}$. The vector $A^k\mathbf{x}_0$ itself may never approach a multiple of $\mathbf{v}_1$ (see Exercises 1 and 21), but if each $A^k\mathbf{x}_0$ is scaled so its largest entry is 1, then the scaled vectors approach an eigenvector (a multiple of $\mathbf{v}_1$) as $k \to \infty$.

The Inverse Power Method: You must start with an initial estimate α for a particular eigenvalue, say λ_2, and α must be closer to λ_2 than to any other eigenvalue of A. In this case, $1/(\lambda_2 - \alpha)$ is a strictly dominant eigenvalue of the matrix $B = (A - \alpha I)^{-1}$. The inverse power method avoids computing B. Instead of multiplying $\mathbf{x}_k$ by B to get $\mathbf{x}_{k+1}$ (suitably scaled), you solve the equation $(A - \alpha I)\mathbf{y}_k = \mathbf{x}_k$ for $\mathbf{y}_k$ and then scale $\mathbf{y}_k$ to produce $\mathbf{x}_{k+1}$.

SOLUTIONS TO EXERCISES

1. The vectors in the sequence $\begin{bmatrix}1\\0\end{bmatrix}, \begin{bmatrix}1\\.25\end{bmatrix}, \begin{bmatrix}1\\.3158\end{bmatrix}, \begin{bmatrix}1\\.3298\end{bmatrix}, \begin{bmatrix}1\\.3326\end{bmatrix}$ approach an eigenvector $\mathbf{v}_1$. Of these vectors, the last one, $\mathbf{x}_4$, is probably the best estimate of $\mathbf{v}_1$. To compute an estimate of λ_1, multiply one of the vectors by A and examine its entries. Again, the best information probably comes from $A\mathbf{x}_4 = \begin{bmatrix}4.9978\\1.6652\end{bmatrix}$, whose entries are approximately λ_1 times the entries in $\mathbf{x}_4$. From the first entry, the estimate of λ_1 is 4.9978.

   The computed value of $A\mathbf{x}_4$ can be used as an estimate of the direction of the eigenspace, but this vector is not necessarily a better estimate for an eigenvector than $\mathbf{x}_4$. For that, you should use $\mathbf{x}_5$, the *scaled* version of $A\mathbf{x}_4$. With the given data, the distance from $A\mathbf{x}_4$ to the eigenspace is about 7.0×10^{-4}, but the distance from $\mathbf{x}_5$ to the eigenspace is only 1.4×10^{-4}.

Study Tip: Exercises 1-6 make good exam questions because they test your understanding of the power method without requiring extensive calculation.

7. The data in the table below and the tables in Exercise 19 were produced by MATLAB, which carried more decimal places than shown here.

k	0	1	2	3	4	5
$\mathbf{x}_k$	$\begin{bmatrix}1\\0\end{bmatrix}$	$\begin{bmatrix}.75\\1\end{bmatrix}$	$\begin{bmatrix}1\\.9565\end{bmatrix}$	$\begin{bmatrix}.9932\\1\end{bmatrix}$	$\begin{bmatrix}1\\.9990\end{bmatrix}$	$\begin{bmatrix}.9998\\1\end{bmatrix}$
$A\mathbf{x}_k$	$\begin{bmatrix}6\\8\end{bmatrix}$	$\begin{bmatrix}11.5\\11.0\end{bmatrix}$	$\begin{bmatrix}12.70\\12.78\end{bmatrix}$	$\begin{bmatrix}12.959\\12.946\end{bmatrix}$	$\begin{bmatrix}12.9927\\12.9948\end{bmatrix}$	$\begin{bmatrix}12.9990\\12.9987\end{bmatrix}$
μ_k	8	11.5	12.78	12.959	12.9948	12.9990

The exact eigenvalues are 13 and -2. The subspaces determined by $A^k\mathbf{x}$ are lines whose slopes alternate above and below the slope of the eigenspace (the eigenspace is the line $x_2 = x_1$).

13. If the eigenvalues close to 4 and -4 have different absolute values, then one of these eigenvalues is a strictly dominant eigenvalue, so the power method will work. But the power method depends on powers of the quotients λ_2/λ_1 and λ_3/λ_1 going to zero. If $|\lambda_2/\lambda_1|$ is close to 1, its powers will go to zero slowly.

15. Suppose $A\mathbf{x} = \lambda\mathbf{x}$, with $\mathbf{x} \neq \mathbf{0}$. For any α, $A\mathbf{x} - \alpha I\mathbf{x} = (\lambda - \alpha)\mathbf{x}$, and $(A - \alpha I)\mathbf{x} = (\lambda - a)\mathbf{x}$. If α is *not* an eigenvalue of A, then $A - \alpha I$ is invertible, and hence

$$\mathbf{x} = (A - \alpha I)^{-1}(\lambda - \alpha)\mathbf{x} \quad \text{and} \quad (\lambda - \alpha)^{-1}\mathbf{x} = (A - \alpha I)^{-1}\mathbf{x}$$

This last equation shows that $\mathbf{x}$ is an eigenvector of $(A - \alpha I)^{-1}$ corresponding to the eigenvalue $(\lambda - \alpha)^{-1}$.

19. a. The data in the table on the next page show that $\mu_6 = 30.2887 = \mu_7$ to four decimal places. Actually, to six places, the largest eigenvalue is 30.288685, with eigenvector (.957629, .688937, 1, .943782).

k	0	1	2	3	4	5	6	7
$\mathbf{x}_k$	$\begin{bmatrix}1\\0\\0\\0\end{bmatrix}$	$\begin{bmatrix}1\\.7\\.8\\.7\end{bmatrix}$	$\begin{bmatrix}.99\\.71\\1\\.93\end{bmatrix}$	$\begin{bmatrix}.961\\.691\\1\\.942\end{bmatrix}$	$\begin{bmatrix}.9581\\.6893\\1\\.9436\end{bmatrix}$	$\begin{bmatrix}.9577\\.6890\\1\\.9438\end{bmatrix}$	$\begin{bmatrix}.957637\\.688942\\1\\.943778\end{bmatrix}$	$\begin{bmatrix}.957630\\.688938\\1\\.943781\end{bmatrix}$
$A\mathbf{x}_k$	$\begin{bmatrix}10\\7\\8\\7\end{bmatrix}$	$\begin{bmatrix}26.2\\18.8\\26.5\\24.7\end{bmatrix}$	$\begin{bmatrix}29.4\\21.1\\30.6\\28.8\end{bmatrix}$	$\begin{bmatrix}29.05\\20.90\\30.32\\28.61\end{bmatrix}$	$\begin{bmatrix}29.01\\20.87\\30.29\\28.59\end{bmatrix}$	$\begin{bmatrix}29.006\\20.868\\30.289\\28.586\end{bmatrix}$	$\begin{bmatrix}29.0054\\20.8671\\30.2887\\29.5859\end{bmatrix}$	$\begin{bmatrix}29.0053\\20.8670\\30.2887\\28.5859\end{bmatrix}$
μ_k	10	26.5	30.6	30.32	30.29	30.2892	30.2887	30.2887

b. The inverse power method (with $\alpha = 0$) produces $\nu_1 = \mu_1^{-1} = .010141$, and $\nu_2 = .0101501$, which seems to be accurate to at least four places. Actually, ν_2 is accurate to six places, ν_3 is accurate to eight places , and ν_4 is accurate to ten places. The convergence is so rapid because the next-to-smallest eigenvalue is near .85, which is much farther away from 0 than .0101500. The vector $\mathbf{x}_4$ gives an estimate for the eigenvector that is accurate to seven places in each entry.

k	0	1	2	3	4
$\mathbf{x}_k$	$\begin{bmatrix}1\\0\\0\\0\end{bmatrix}$	$\begin{bmatrix}-.6098\\1\\-.2439\\.1463\end{bmatrix}$	$\begin{bmatrix}-.60401\\1\\-.25105\\.14890\end{bmatrix}$	$\begin{bmatrix}-.603973\\1\\-.251134\\.148953\end{bmatrix}$	$\begin{bmatrix}-.6039723\\1\\-.2511351\\.1489534\end{bmatrix}$
$\mathbf{y}_k$	$\begin{bmatrix}25\\-41\\10\\-6\end{bmatrix}$	$\begin{bmatrix}-59.56\\98.61\\-24.76\\14.68\end{bmatrix}$	$\begin{bmatrix}-59.5041\\98.5211\\-24.7420\\14.6750\end{bmatrix}$	$\begin{bmatrix}-59.5044\\98.5217\\-24.7423\\14.6751\end{bmatrix}$	$\begin{bmatrix}-59.50438\\98.52170\\-24.74226\\14.67515\end{bmatrix}$
ν_k	-.024	.010141	.0101501	.010150049	.0101500484

MATLAB Use **format long** to display 15 decimal digits in your data. The algorithms below assume that A has a strictly dominant eigenvalue, and the initial vector is **x**, with largest entry 1 (in magnitude). (If your initial vector is called **x0**, rename it by entering **x = x0** .)

The Power Method When the following commands are repeated over and over, the values of **x** approach (in many cases) an eigenvector for a strictly dominant eigenvalue:

```
y = A*x                                                         (1)
[t  r] = max(abs(y)); mu = y(r)    mu = estimate for eigenvalue (2)
x = y/y(r)                         Estimate for the eigenvector (3)
```

In (2), t is the absolute value of the largest entry in **y** and r is the index of that entry. As these commands are repeated, the numbers that appear in **y**(r) are the μ_k that approach the dominant eigenvalue.

Recall that MATLAB commands can be recalled by the up-arrow key (↑). After entering (1)-(3), your keystrokes can be

↑ ↑ ↑ <Enter> ↑ ↑ ↑ <Enter> ↑ ↑ ↑ <Enter>

and so on. Alternatively, you could enclose lines (1)-(3) in a loop (see page 6-25).

The Inverse Power Method Store the initial estimate of the eigenvalue in the variable **a**, and enter the command **C = A - a*eye(n)**, where n is the number of columns of A. Then enter the commands

```
y = C\x                        Solves the equation (A - aI)y = x   (1)
[t  r] = max(abs(y)); nu = a + 1/y(r)   nu = estimated e-value     (2)
x = y/y(r)                     Estimate for the eigenvector        (3)
```

As these commands are repeated (using ↑↑↑ <Enter> each time), lines (2) and (3) produce the sequences $\{\nu_k\}$ and $\{\mathbf{x}_k\}$ described in the text.

Displaying Data If your computer screen displays only 24 or 25 lines, vectors in the sequence $\{\mathbf{x}_k\}$ tend to scroll off the screen soon after you compute them. To see more vectors at once, and to compare their entries more easily, you can *display* them as row vectors. Change (1) to **y = A*x; y'** (power method) or **y = C\x; y'** (inverse power), and for both methods, change (3) to **x = y/y(r); x'** .

For even more data on your screen, use the command **format compact** , which removes extra lines between data displays. The simple command **format** returns everything to normal.

CHAPTER 6 GLOSSARY CHECKLIST

Check your knowledge by attempting to write definitions of the terms below. Then compare your work with the definitions given in the text's Glossary. Ask your instructor which definitions, if any, might appear on a test.

algebraic multiplicity: The multiplicity of an eigenvalue as ....

attractor (of a dynamical system $\mathbf{x}_{k+1} = A\mathbf{x}_k$): The origin of $\mathbb{R}^n$ when .... All trajectories tend ....

$\mathcal{B}$-matrix (for T): A matrix $[T]_{\mathcal{B}}$ for a linear transformation $T: V \to V$ relative to a basis $\mathcal{B}$ for V, with the property that ....

characteristic equation (of A): ....

characteristic polynomial (of A): ....

companion matrix: A special form of matrix whose characteristic ... is ....

complex eigenvalue: A nonreal root of the characteristic equation of an $n \times n$ matrix A, when ....

complex eigenvector: A nonzero vector $\mathbf{x}$ in $\mathbb{C}^n$ such that ..., where ....

decoupled system: A difference equation $\mathbf{y}_{k+1} = A\mathbf{y}_k$ where ....

determinant (of a square matrix A): A number $\det A$ computed from A; equal to ....

diagonalizable (matrix): A matrix that may be written in ... form as ....

difference equation (or **linear recurrence relation**): An equation of the form ... whose solution is ....

discrete linear dynamical system (or briefly, a **dynamical system**): A difference equation of the form ... that describes ....

eigenspace (of A corresponding to λ): The set of ... solutions of ....

eigenvalue (of A): A scalar λ such that ....

eigenvector (of A): A ... vector $\mathbf{x}$ such that ....

eigenvector basis: A basis consisting entirely of ....

eigenvector decomposition (of $\mathbf{x}$): An equation $\mathbf{x} = $ ....

Im x: The vector in $\mathbb{R}^n$ formed from ....

invariant subspace (for A): A subspace H such that ....

inverse power method: An algorithm for estimating ....

matrix for T relative to bases $\mathcal{B}$ and $\mathcal{C}$: A matrix M for a linear transformation $T: V \to W$ with the property that .... When $W = V$ and $\mathcal{C} = \mathcal{B}$, the matrix M is called ... and is denoted by ....

power method: An algorithm for estimating ....

repellor (of a dynamical system $\mathbf{x}_{k+1} = A\mathbf{x}_k$) : The origin in $\mathbb{R}^n$ when .... All trajectories ... tend ......

Rayleigh quotient: $R(\mathbf{x})$ = ... An estimate of ....

Re x: The vector in $\mathbb{R}^n$ formed from ....

saddle point (of a dynamical system $\mathbf{x}_{k+1} = A\mathbf{x}_k$): The origin in $\mathbb{R}^n$ when .... Some points $\mathbf{x}$ ....

similar (matrices): Matrices A and B such that ....

similarity transformation: A transformation of a matrix, $A \mapsto$ ....

stage-matrix model: A difference equation $\mathbf{x}_{k+1} = A\mathbf{x}_k$ where $\mathbf{x}_k$ lists ....

strictly dominant eigenvalue: An eigenvalue λ_1 of a matrix A with the property that ...

trace (of a square matrix A): The ..., denoted by $\operatorname{tr} A$.

trajectory: The graph of a solution $\{\mathbf{x}_0, \mathbf{x}_1, \mathbf{x}_2, \ldots\}$ of a ....

7

Orthogonality and Least-Squares

7.1 INNER PRODUCT, LENGTH, AND ORTHOGONALITY

The concepts of length, distance, and orthogonality introduced in this section are essential for many geometric descriptions in the rest of the text.

KEY IDEAS

The first half of the section is computational and easily learned. The second half, however, requires more attention. Read it carefully. The concepts of orthogonality and orthogonal complements are the foundation for the rest of the chapter. In fact, Theorem 3 is sometimes called the Fundamental Theorem of Linear Algebra.

In Figure 8, when a subspace such as Col *A* is referred to as a fundamental subspace of *A*, the phrase "of *A*" points to the important connection between Col *A* and the matrix; the phrase does not imply that Col *A* in some sense lies inside the matrix.

SOLUTIONS TO EXERCISES

1. $\mathbf{u} = \begin{bmatrix} -1 \\ 2 \end{bmatrix}$, $\mathbf{v} = \begin{bmatrix} 4 \\ 6 \end{bmatrix}$, $\mathbf{u}\cdot\mathbf{u} = (-1)^2 + 2^2 = 5$,
$\mathbf{v}\cdot\mathbf{u} = 4(-1) + 6(2) = 8$, $\dfrac{\mathbf{v}\cdot\mathbf{u}}{\mathbf{u}\cdot\mathbf{u}} = \dfrac{8}{5}$

7. $\mathbf{u} = \begin{bmatrix} -1 \\ 2 \end{bmatrix}$, $\|\mathbf{u}\|^2 = \mathbf{u}\cdot\mathbf{u} = 5$, from Exercise 1. So $\|\mathbf{u}\| = \sqrt{5}$.

13. $\mathbf{u} = \begin{bmatrix} 7/4 \\ 1/2 \\ 1 \end{bmatrix}$. As in Example 3, first scale $\mathbf{u}$. Let $\mathbf{y} = 4\mathbf{u} = \begin{bmatrix} 7 \\ 2 \\ 4 \end{bmatrix}$. Then $\|\mathbf{y}\|^2 = 7^2 + 2^2 + 4^2 = 69$, $\|\mathbf{y}\| = \sqrt{69}$. A unit vector in the direction of both **u** and **y** is $(1/\sqrt{69})\mathbf{y} = \begin{bmatrix} 7/\sqrt{69} \\ 2/\sqrt{69} \\ 4/\sqrt{69} \end{bmatrix}$.

17. Theorem 1(b): $(\mathbf{u} + \mathbf{v})\cdot\mathbf{w} = (\mathbf{u} + \mathbf{v})^T\mathbf{w} = (\mathbf{u}^T + \mathbf{v}^T)\mathbf{w} = \mathbf{u}^T\mathbf{w} + \mathbf{v}^T\mathbf{w} = \mathbf{u}\cdot\mathbf{w} + \mathbf{v}\cdot\mathbf{w}$. The second and third equalities used Theorems 3(b) and 2(c), respectively, from Section 3.1.

Theorem 1(c):**c**. $(c\mathbf{u})\cdot\mathbf{v} = (c\mathbf{u})^T\mathbf{v} = (c\mathbf{u}^T)\mathbf{v} = c(\mathbf{u}^T\mathbf{v}) = c(\mathbf{u}\cdot\mathbf{v})$, by Theorem 3(c) in Section 3.1. Also, $\mathbf{u}\cdot(c\mathbf{v}) = \mathbf{u}^T(c\mathbf{v}) = c\mathbf{u}^T\mathbf{v} = c(\mathbf{u}\cdot\mathbf{v})$.

19. $\mathbf{a} = \begin{bmatrix} 8 \\ -5 \end{bmatrix}$, $\mathbf{b} = \begin{bmatrix} -2 \\ -3 \end{bmatrix}$, $\mathbf{a}\cdot\mathbf{b} = 8(-2) + (-5)(-3) = -1 \neq 0$, so **a** and **b** are not orthogonal.

27. If **y** is orthogonal to **u** and **v**, then $\mathbf{y}\cdot\mathbf{u} = 0$ and $\mathbf{y}\cdot\mathbf{v} = 0$, and hence by a property of the inner product, $\mathbf{y}\cdot(\mathbf{u} + \mathbf{v}) = \mathbf{y}\cdot\mathbf{u} + \mathbf{y}\cdot\mathbf{v} = 0 + 0 = 0$. So **y** is orthogonal to **u** + **v**.

29. Take a typical vector $\mathbf{w} = c_1\mathbf{v}_1 + \cdots + c_p\mathbf{v}_p$ in W. If **x** is orthogonal to each $\mathbf{v}_j$, then using the linearity of the inner product (Theorem 1(b) and 1(c)),

$$\mathbf{w}\cdot\mathbf{x} = (c_1\mathbf{v}_1 + \cdots + c_p\mathbf{v}_p)\cdot\mathbf{x} = c_1\mathbf{v}_1\cdot\mathbf{x} + \cdots + c_p\mathbf{v}_p\cdot\mathbf{x} = 0$$

So **x** is orthogonal to each **w** in W.

31. Suppose **x** is in W and $W^\perp$. Then, since **x** is in $W^\perp$, **x** is orthogonal to every vector in W, including **x** itself. So $\mathbf{x}\cdot\mathbf{x} = 0$. This is true only if $\mathbf{x} = \mathbf{0}$. This problem shows that $W \cap W^\perp$ is the zero subspace.

MATLAB The inner product of column vectors **u** and **v** is **u'*v** (and **v'*u**); the length of **v** is **norm(v)** . See the MATLAB note on page 3-4.

7.2 ORTHOGONAL SETS

Orthogonal sets and orthogonal bases are used throughout the chapter. The "orthogonal projection" discussed in this section is an important special case of the orthogonal projections studied in Section 7.3.

STUDY NOTES

The proofs of Theorems 4 and 5 are worth studying because they involve a calculation you will see and use several times.

The attention paid to Theorems 6 and 7 will depend on what your instructor plans to do later in the course. In some cases, an instructor may discuss Theorems 6 and 7 only for square matrices. The $m \times n$ case is needed later, for Theorems 10, 12 and 15. Remember: the term *orthogonal matrix* applies only to a square matrix. Also, the columns of an orthogonal matrix must be ortho*normal*, not simply orthogonal.

The subsection entitled "An Orthogonal Projection" is simple but extremely important. Finally, the geometric interpretation of Theorem 5, on page 349, will be helpful when you study Theorem 8 in the next section.

SOLUTIONS TO EXERCISES

1. $\mathbf{u} = \begin{bmatrix} -1 \\ 4 \\ -3 \end{bmatrix}$, $\mathbf{v} = \begin{bmatrix} 5 \\ 2 \\ 1 \end{bmatrix}$, $\mathbf{w} = \begin{bmatrix} 3 \\ -4 \\ -7 \end{bmatrix}$, $\mathbf{u}\cdot\mathbf{v} = -5 + 8 - 3 = 0$,

$\mathbf{u}\cdot\mathbf{w} = -3 - 16 + 21 = 2 \neq 0$. The set $\{\mathbf{u}, \mathbf{v}, \mathbf{w}\}$ is not orthogonal. There is no need to check $\mathbf{v}\cdot\mathbf{w}$.

7. $\mathbf{u}_1 = \begin{bmatrix} 2 \\ -3 \end{bmatrix}$, $\mathbf{u}_2 = \begin{bmatrix} 6 \\ 4 \end{bmatrix}$, $\mathbf{x} = \begin{bmatrix} 9 \\ -7 \end{bmatrix}$. $\mathbf{u}_1\cdot\mathbf{u}_2 = 12 - 12 = 0$, so $\{\mathbf{u}_1, \mathbf{u}_2\}$ is an orthogonal set. Since the vectors are nonzero, $\mathbf{u}_1$ and $\mathbf{u}_2$ are linearly independent, by Theorem 4. But two such vectors in $\mathbb{R}^2$ automatically form a basis for $\mathbb{R}^2$. So $\{\mathbf{u}_1, \mathbf{u}_2\}$ is an orthogonal basis for $\mathbb{R}^2$. By Theorem 5,

$$\mathbf{x} = \frac{\mathbf{x}\cdot\mathbf{u}_1}{\mathbf{u}_1\cdot\mathbf{u}_1}\mathbf{u}_1 + \frac{\mathbf{x}\cdot\mathbf{u}_2}{\mathbf{u}_2\cdot\mathbf{u}_2}\mathbf{u}_2 = \frac{18 + 21}{4 + 9}\mathbf{u}_1 + \frac{54 - 28}{36 + 16}\mathbf{u}_2 = 3\begin{bmatrix} 2 \\ -3 \end{bmatrix} + \frac{1}{2}\begin{bmatrix} 6 \\ 4 \end{bmatrix}$$

13. $\mathbf{u} = \begin{bmatrix} -.6 \\ .8 \end{bmatrix}$, $\mathbf{v} = \begin{bmatrix} .8 \\ .6 \end{bmatrix}$, $\mathbf{u}\cdot\mathbf{v} = -.48 + .48 = 0$, so $\{\mathbf{u}, \mathbf{v}\}$ is an orthogonal set. Also, $\|\mathbf{u}\|^2 = \mathbf{u}\cdot\mathbf{u} = (-.6)^2 + (.8)^2 = .36 + .64 = 1$. Similarly, $\|\mathbf{v}\|^2 = \mathbf{v}\cdot\mathbf{v} = 1$. Thus $\{\mathbf{u}, \mathbf{v}\}$ is an orthonormal set.

19. $\mathbf{y} = \begin{bmatrix} 2 \\ 3 \end{bmatrix}$, $\mathbf{u} = \begin{bmatrix} 4 \\ -7 \end{bmatrix}$. The orthogonal projection of $\mathbf{y}$ onto $\mathbf{u}$ is

$$\hat{\mathbf{y}} = \frac{\mathbf{y}\cdot\mathbf{u}}{\mathbf{u}\cdot\mathbf{u}}\mathbf{u} = \frac{8 - 21}{16 + 49}\mathbf{u} = \frac{-13}{65}\mathbf{u} = \frac{-1}{5}\begin{bmatrix} 4 \\ -7 \end{bmatrix} = \begin{bmatrix} -4/5 \\ 7/5 \end{bmatrix}$$

The component of **y** orthogonal to **u** is $\mathbf{y} - \hat{\mathbf{y}} = \begin{bmatrix} 2 \\ 3 \end{bmatrix} - \begin{bmatrix} -4/5 \\ 7/5 \end{bmatrix} = \begin{bmatrix} 14/5 \\ 8/5 \end{bmatrix}$.

Thus, $\mathbf{y} = \hat{\mathbf{y}} + (\mathbf{y} - \hat{\mathbf{y}}) = \begin{bmatrix} -4/5 \\ 7/5 \end{bmatrix} + \begin{bmatrix} 14/5 \\ 8/5 \end{bmatrix}$.

25. $\|U\mathbf{x}\|^2 = (U\mathbf{x})^T(U\mathbf{x}) = \mathbf{x}^T U^T U\mathbf{x} = \mathbf{x}^T\mathbf{x} = \|\mathbf{x}\|^2$ (using Theorem 6).

Study Tip: If your instructor emphasizes orthogonal matrices, you should work Exercises 27—29. (They make good test questions.) In each case, mention explicitly how you use the fact that the matrices are square. Don't read the solutions below until you have first *written* down your own solution.

27. If U has orthonormal columns, then $U^TU = I$, by Theorem 6. If U is also *square*, then the equation $U^TU = I$ implies that U is invertible, by the Invertible Matrix Theorem.

29. Since U and V are orthogonal, each is invertible. By Theorem 6 in Section 3.2, UV is invertible and $(UV)^{-1} = V^{-1}U^{-1} = V^TU^T = (UV)^T$ (by Theorem 3 in Section 3.1). Thus UV is an orthogonal matrix.

MATLAB In Exercises 1—16, the fastest way (counting the keystrokes) in MATLAB to test a set such as $\{\mathbf{u}_1, \mathbf{u}_2, \mathbf{u}_3\}$ for orthogonality is to use a matrix **U = [u1 u2 u3]** whose columns are the vectors from the set. See the proof of Theorem 6.

For column vectors **y** and **u**, the orthogonal projection of **y** onto **u** is

```
(y'*u)/(u'*u)*u
```

The parentheses (and the final *) are essential. MATLAB computes the scalar quotient $(\mathbf{y}^T\mathbf{u})/(\mathbf{u}^T\mathbf{u})$ and then multiplies **u** by this scalar.

7.3 ORTHOGONAL PROJECTIONS

A familiar idea in Euclidean geometry is to construct a line segment perpendicular to a line or plane. This section treats an analogous situation in $\mathbb{R}^n$, namely, the orthogonal projection of a vector (a point in $\mathbb{R}^n$) onto a subspace. The case when the subspace is a line through the origin was already examined in Section 7.2.

KEY IDEAS

If $\mathbf{y}$ is in $\mathbb{R}^n$ and if W is a subspace of $\mathbb{R}^n$, then the orthogonal projection of $\mathbf{y}$ onto W, denoted by $\hat{\mathbf{y}}$ or $\text{proj}_W\,\mathbf{y}$, has two important properties:

(i) $\mathbf{y} - \hat{\mathbf{y}}$ is orthogonal to W (so $\mathbf{y}$ is the sum of a vector $\hat{\mathbf{y}}$ in W and a vector $\mathbf{y} - \hat{\mathbf{y}}$ in $W^\perp$), and

(ii) $\hat{\mathbf{y}}$ is the closest point in W to $\mathbf{y}$.

Properties (i) and (ii) are described in the Orthogonal Decomposition Theorem and the Best Approximation Theorem. You should learn the statements of both theorems. (By now you probably know that whenever a theorem has an official name, an instructor has an easy time asking test questions about it.) When you need one of these theorems in a discussion (homework or test question), you should mention the theorem by name.

If your class covers Theorem 10, then the paragraph following the theorem will help you understand the difference between an *orthogonal matrix* (which must be square) and a rectangular matrix with orthonormal columns.

SOLUTIONS TO EXERCISES

1. $\mathbf{u}_1 = \begin{bmatrix} 0 \\ 1 \\ -4 \\ -1 \end{bmatrix}$, $\mathbf{u}_2 = \begin{bmatrix} 3 \\ 5 \\ 1 \\ 1 \end{bmatrix}$, $\mathbf{u}_3 = \begin{bmatrix} 1 \\ 0 \\ 1 \\ -4 \end{bmatrix}$, $\mathbf{u}_4 = \begin{bmatrix} 5 \\ -3 \\ -1 \\ 1 \end{bmatrix}$, $\mathbf{x} = \begin{bmatrix} 10 \\ -8 \\ 2 \\ 0 \end{bmatrix}$. You could calculate all the inner products in the decomposition:

$$\mathbf{x} = \underbrace{\frac{\mathbf{x}\cdot\mathbf{u}_1}{\mathbf{u}_1\cdot\mathbf{u}_1}\mathbf{u}_1 + \frac{\mathbf{x}\cdot\mathbf{u}_2}{\mathbf{u}_2\cdot\mathbf{u}_2}\mathbf{u}_2 + \frac{\mathbf{x}\cdot\mathbf{u}_3}{\mathbf{u}_3\cdot\mathbf{u}_3}\mathbf{u}_3}_{\text{in Span}\{\mathbf{u}_1,\mathbf{u}_2,\mathbf{u}_3\}} + \underbrace{\frac{\mathbf{x}\cdot\mathbf{u}_4}{\mathbf{u}_4\cdot\mathbf{u}_4}\mathbf{u}_4}_{\text{in Span}\{\mathbf{u}_4\}} \qquad (*)$$

However, once you know the vector in Span$\{\mathbf{u}_4\}$, the vector in Span$\{\mathbf{u}_1, \mathbf{u}_2, \mathbf{u}_3\}$ is determined completely by (*). So all you need is

$$\frac{\mathbf{x}\cdot\mathbf{u}_4}{\mathbf{u}_4\cdot\mathbf{u}_4}\mathbf{u}_4 = \frac{50 + 24 - 2 + 0}{25 + 9 + 1 + 1}\mathbf{u}_4 = 2\mathbf{u}_4 = \begin{bmatrix} 10 \\ -6 \\ -2 \\ 2 \end{bmatrix}$$

The vector in Span$\{\mathbf{u}_1, \mathbf{u}_2, \mathbf{u}_3\}$ is $\mathbf{x} - 2\mathbf{u}_4 = \begin{bmatrix} 10 \\ -8 \\ 2 \\ 0 \end{bmatrix} - \begin{bmatrix} 10 \\ -6 \\ -2 \\ 2 \end{bmatrix} = \begin{bmatrix} 0 \\ -2 \\ 4 \\ -2 \end{bmatrix}$.

Study Tip: One way to check whether $\text{proj}_W \mathbf{y}$ is computed correctly is to verify that $\mathbf{y} - \text{proj}_W \mathbf{y}$ is orthogonal to each vector in the orthogonal basis $\{\mathbf{u}_1, \ldots, \mathbf{u}_p\}$ for W. A faster check that will catch most errors (but not all) is to verify that $\mathbf{y} - \text{proj}_W \mathbf{y}$ is orthogonal to $\text{proj}_W \mathbf{y}$.

7. $\mathbf{y} = \begin{bmatrix} 1 \\ 3 \\ 5 \end{bmatrix}$, $\mathbf{u}_1 = \begin{bmatrix} 1 \\ 3 \\ -2 \end{bmatrix}$, $\mathbf{u}_2 = \begin{bmatrix} 5 \\ 1 \\ 4 \end{bmatrix}$. First, make sure that $\{\mathbf{u}_1, \mathbf{u}_2\}$ is an orthogonal basis for $\text{Span}\{\mathbf{u}_1, \mathbf{u}_2\}$. This is easy, since $\mathbf{u}_1$ and $\mathbf{u}_2$ are nonzero and $\mathbf{u}_1 \cdot \mathbf{u}_2 = 0$. Next, by the Orthogonal Decomposition Theorem, $\mathbf{y}$ is the sum of $\text{proj}_W \mathbf{y}$ and $\mathbf{y} - \text{proj}_W \mathbf{y}$, where $W = \text{Span}\{\mathbf{u}_1, \mathbf{u}_2\}$.

$$\text{proj}_W \mathbf{y} = \frac{\mathbf{y} \cdot \mathbf{u}_1}{\mathbf{u}_1 \cdot \mathbf{u}_1}\mathbf{u}_1 + \frac{\mathbf{y} \cdot \mathbf{u}_2}{\mathbf{u}_2 \cdot \mathbf{u}_2}\mathbf{u}_2 = \frac{1 + 9 - 10}{1 + 9 + 4}\mathbf{u}_1 + \frac{5 + 3 + 20}{25 + 1 + 16}\mathbf{u}_2$$

$$= \mathbf{0} + \frac{28}{42}\begin{bmatrix} 5 \\ 1 \\ 4 \end{bmatrix} = \begin{bmatrix} 10/3 \\ 2/3 \\ 8/3 \end{bmatrix}$$

and

$$\mathbf{y} - \text{proj}_W \mathbf{y} = \begin{bmatrix} 1 \\ 3 \\ 5 \end{bmatrix} - \begin{bmatrix} 10/3 \\ 2/3 \\ 8/3 \end{bmatrix} = \begin{bmatrix} -7/3 \\ 7/3 \\ 7/3 \end{bmatrix}$$

As a check, scale $\mathbf{y} - \text{proj}_W \mathbf{y}$ to $\begin{bmatrix} -1 \\ 1 \\ 1 \end{bmatrix}$, and observe that the scaled vector is obviously orthogonal to $\mathbf{u}_1$ and $\mathbf{u}_2$. Thus $\mathbf{y} - \text{proj}_W \mathbf{y}$ is in $W^{\perp}$, as it should be.

Warning: The formula for $\text{proj}_W \mathbf{y}$ applies only if $\{\mathbf{u}_1, \ldots, \mathbf{u}_p\}$ is an *orthogonal* basis for W. That's why you should check orthogonality, as in Exercise 7, if you are not sure that the basis is orthogonal. If an orthogonal basis is not available, then other methods can be used to compute $\hat{\mathbf{y}}$. (See Exercise 21 in Section 7.5, for example.)

13. $\mathbf{z} = \begin{bmatrix} 3 \\ -7 \\ 2 \\ 3 \end{bmatrix}$, $\mathbf{v}_1 = \begin{bmatrix} 2 \\ -1 \\ -3 \\ 1 \end{bmatrix}$, $\mathbf{v}_2 = \begin{bmatrix} 1 \\ 1 \\ 0 \\ -1 \end{bmatrix}$. Note that $\mathbf{v}_1$ and $\mathbf{v}_2$ are orthogonal. By the Best Approximation Theorem, the closest point in $\text{Span}\{\mathbf{v}_1, \mathbf{v}_2\}$ to $\mathbf{z}$ is the orthogonal projection $\hat{\mathbf{z}}$, where

$$\hat{\mathbf{z}} = \frac{\mathbf{z}\cdot\mathbf{v}_1}{\mathbf{v}_1\cdot\mathbf{v}_1}\mathbf{v}_1 + \frac{\mathbf{z}\cdot\mathbf{v}_2}{\mathbf{v}_2\cdot\mathbf{v}_2}\mathbf{v}_2 = \frac{10}{15}\mathbf{v}_1 + \frac{-7}{3}\mathbf{v}_2 = \frac{2}{3}\begin{bmatrix}2\\-1\\-3\\1\end{bmatrix} - \frac{7}{3}\begin{bmatrix}1\\1\\0\\-1\end{bmatrix} = \begin{bmatrix}-1\\-3\\-2\\3\end{bmatrix}$$

Check: $\mathbf{z} - \hat{\mathbf{z}} = \begin{bmatrix}4\\-4\\4\\0\end{bmatrix}$. The vector $\begin{bmatrix}1\\-1\\1\\0\end{bmatrix}$ is orthogonal to both $\mathbf{v}_1$ and $\mathbf{v}_2$.

19. By the Orthogonal Decomposition Theorem, $\mathbf{u}_3$ is the sum of a vector in $W = \text{Span}\{\mathbf{u}_1, \mathbf{u}_2\}$ and a vector $\mathbf{v}$ orthogonal to W. First,

$$\text{proj}_W\, \mathbf{u}_3 = \frac{-2}{6}\mathbf{u}_1 + \frac{2}{30}\mathbf{u}_2 = \begin{bmatrix}-2/6\\-2/6\\4/6\end{bmatrix} + \begin{bmatrix}10/30\\-2/30\\4/30\end{bmatrix} = \begin{bmatrix}0\\-2/5\\4/5\end{bmatrix}$$

Then

$$\mathbf{v} = \mathbf{u}_3 - \text{proj}_W\, \mathbf{u}_3 = \begin{bmatrix}0\\0\\1\end{bmatrix} - \begin{bmatrix}0\\-2/5\\4/5\end{bmatrix} = \begin{bmatrix}0\\2/5\\1/5\end{bmatrix}$$

Not only is $\mathbf{v}$ orthogonal to W, but also any multiple of $\mathbf{v}$ is in $W^\perp$.

Study Tip: It would be a good idea to try Exercise 20 and compare the result with Exercise 19. Then think about the following problem: Suppose that $\{\mathbf{u}_1, \mathbf{u}_2\}$ is an orthogonal set of nonzero vectors in $\mathbb{R}^3$. How would you find an orthogonal basis of $\mathbb{R}^3$ that contains $\mathbf{u}_1$ and $\mathbf{u}_2$? You might discuss this with your instructor.

21. By the Orthogonal Decomposition Theorem, each $\mathbf{x}$ in $\mathbb{R}^n$ can be written uniquely as $\mathbf{x} = \mathbf{p} + \mathbf{u}$, with $\mathbf{p}$ in Row A and $\mathbf{u}$ in $(\text{Row}\, A)^\perp$. By Theorem 3 in Section 7.1, $\mathbf{u}$ is in Nul A.

Next, suppose that $A\mathbf{x} = \mathbf{b}$ is consistent. Let $\mathbf{x}$ be a solution, and write $\mathbf{x} = \mathbf{p} + \mathbf{u}$, as above. Then $A\mathbf{p} = A(\mathbf{x} - \mathbf{u}) = A\mathbf{x} - A\mathbf{u} = \mathbf{b} - \mathbf{0} = \mathbf{b}$. So the equation $A\mathbf{x} = \mathbf{b}$ has at least one solution $\mathbf{p}$ in Row A.

Finally, suppose that $\mathbf{p}$ and $\mathbf{p}_1$ are both in Row A and satisfy $A\mathbf{x} = \mathbf{b}$. Then $\mathbf{p} - \mathbf{p}_1$ is in Nul A because

$$A(\mathbf{p} - \mathbf{p}_1) = A\mathbf{p} - A\mathbf{p}_1 = \mathbf{b} - \mathbf{b} = \mathbf{0}$$

The equations $\mathbf{p} = \mathbf{p}_1 + (\mathbf{p} - \mathbf{p}_1)$ and $\mathbf{p} = \mathbf{p} + \mathbf{0}$ both decompose $\mathbf{p}$ as the sum of a vector in Row A and a vector in $(\text{Row}\, A)^\perp$. By the uniqueness of the orthogonal decomposition (Theorem 8), $\mathbf{p}_1 = \mathbf{p}$, so $\mathbf{p}$ is unique.

Warning: I had to work hard to make the arithmetic simple in the exercises for this section, to avoid distractions for you and to save you time. You might not be so lucky on an exam. Even if a problem is designed to be numerically simple, there is always a chance that a minor error will make the calculations messy. In such a case, don't despair. Carry out the arithmetic as best you can, showing the details of your work (patterned after the solutions in this Study Guide). Chances are that you will get substantial credit for showing that you understand the concepts.

MATLAB The orthogonal projection of **y** onto a single vector was described in the MATLAB note for Section 7.2. The orthogonal projection onto the set spanned by an orthogonal set of vectors is the sum of the one-dimensional projections. Another way to construct this projection is to normalize the orthogonal vectors, place them in the columns of a matrix U, and use Theorem 10. For instance, if $\{\mathbf{y}_1, \mathbf{y}_2, \mathbf{y}_3\}$ is an orthogonal set of nonzero vectors, then the matrix

```
U = [y1/norm(y1)  y2/norm(y2)  y3/norm(y3)]
```

has orthonormal columns, and **U*(U'*y)** produces the orthogonal projection of **y** onto the subspace spanned by $\{\mathbf{y}_1, \mathbf{y}_2, \mathbf{y}_3\}$. (The parentheses around **U'*y** speeds up the computation of **U*U'*y**.)

7.4 THE GRAM-SCHMIDT PROCESS

This section has a nice geometric appeal. The Gram-Schmidt process is well-liked by students and faculty because it is easily learned. Although the process is seldom used in practical work, it has important generalizations to spaces other than $\mathbb{R}^n$ (to be discussed briefly in Section 7.7).

KEY IDEAS

When the Gram-Schmidt process is applied to $\{\mathbf{x}_1, \ldots, \mathbf{x}_p\}$, the first step is to set $\mathbf{v}_1 = \mathbf{x}_1$. For $k = 2, \ldots, n$, the kth step consists of subtracting from $\mathbf{x}_k$ its projection onto the subspace spanned by the previous **x**'*s*. At each step the projection is easy to compute because an orthogonal basis for the appropriate subspace has already been constructed.

The QR factorization of a matrix A encapsulates the result of applying the Gram-Schmidt process to the columns of A, just as the LU factorization of a matrix encodes the row operations that reduce a matrix to echelon form. Also, just as the LU factorization can be implemented via multiplication by elementary matrices, so can the QR factorization be constructed via multiplication by orthogonal matrices.

SOLUTIONS TO EXERCISES

1. $\mathbf{x}_1 = \begin{bmatrix} 3 \\ 0 \\ -1 \end{bmatrix}$, $\mathbf{x}_2 = \begin{bmatrix} 8 \\ 5 \\ -6 \end{bmatrix}$. Set $\mathbf{v}_1 = \mathbf{x}_1$ and compute

$$\mathbf{v}_2 = \mathbf{x}_2 - \frac{\mathbf{x}_2 \cdot \mathbf{v}_1}{\mathbf{v}_1 \cdot \mathbf{v}_1}\mathbf{v}_1 = \begin{bmatrix} 8 \\ 5 \\ -6 \end{bmatrix} - \frac{30}{10}\begin{bmatrix} 3 \\ 0 \\ -1 \end{bmatrix} = \begin{bmatrix} -1 \\ 5 \\ -3 \end{bmatrix}$$

Check: $\mathbf{v}_2 \cdot \mathbf{v}_1 = -3 + 0 + 3 = 0$. So an orthogonal basis is $\left\{ \begin{bmatrix} 3 \\ 0 \\ -1 \end{bmatrix}, \begin{bmatrix} -1 \\ 5 \\ -3 \end{bmatrix} \right\}$.

7. $\mathbf{x}_1 = \begin{bmatrix} 2 \\ -5 \\ 1 \end{bmatrix}$, $\mathbf{x}_2 = \begin{bmatrix} 4 \\ -1 \\ 2 \end{bmatrix}$. From Exercise 3, use $\mathbf{v}_1 = \begin{bmatrix} 2 \\ -5 \\ 1 \end{bmatrix}$ and $\mathbf{v}_2 = \begin{bmatrix} 3 \\ 3/2 \\ 3/2 \end{bmatrix}$ as an orthogonal basis for $W = \text{Span}\{\mathbf{x}_1, \mathbf{x}_2\}$. Scale $\mathbf{v}_2$ to (2, 1, 1) before normalizing, and then obtain

$$\mathbf{u}_1 = \frac{1}{\sqrt{30}}\begin{bmatrix} 2 \\ -5 \\ 1 \end{bmatrix} = \begin{bmatrix} 2/\sqrt{30} \\ -5/\sqrt{30} \\ 1/\sqrt{30} \end{bmatrix}, \quad \mathbf{u}_2 = \frac{1}{\sqrt{6}}\begin{bmatrix} 2 \\ 1 \\ 1 \end{bmatrix} = \begin{bmatrix} 2/\sqrt{6} \\ 1/\sqrt{6} \\ 1/\sqrt{6} \end{bmatrix}$$

Study Tip: If you need to normalize a vector by hand, first consider scaling the entries in the vector to make them small integers, if possible.

13. $A = \begin{bmatrix} 5 & 9 \\ 1 & 7 \\ -3 & -5 \\ 1 & 5 \end{bmatrix}$, $Q = \begin{bmatrix} 5/6 & -1/6 \\ 1/6 & 5/6 \\ -3/6 & 1/6 \\ 1/6 & 3/6 \end{bmatrix}$. Let

$$R = Q^T A = \begin{bmatrix} 5/6 & 1/6 & -3/6 & 1/6 \\ -1/6 & 5/6 & 1/6 & 3/6 \end{bmatrix}\begin{bmatrix} 5 & 9 \\ 1 & 7 \\ -3 & -5 \\ 1 & 5 \end{bmatrix} = \begin{bmatrix} 36/6 & 72/6 \\ 0 & 36/6 \end{bmatrix} = \begin{bmatrix} 6 & 12 \\ 0 & 6 \end{bmatrix}$$

As a check, compute $QR = \begin{bmatrix} 5/6 & -1/6 \\ 1/6 & 5/6 \\ -3/6 & 1/6 \\ 1/6 & 3/6 \end{bmatrix}\begin{bmatrix} 6 & 12 \\ 0 & 6 \end{bmatrix} = \begin{bmatrix} 5 & 54/6 \\ 1 & 42/6 \\ -3 & -30/6 \\ 1 & 30/6 \end{bmatrix} = A.$

Remark: The reason the R in Exercise 13 works is that the columns of Q form an orthonormal basis for Col A (since they were obtained by the Gram-Schmidt process). Thus $QQ^T\mathbf{y} = \mathbf{y}$ for all $\mathbf{y}$ in Col A, by Theorem 10 in Section 7.3. In particular, $QQ^TA = A$. So if R is Q^TA, then $QR = Q(Q^TA) = A$.

19. The solution in the text is complete, except for the details of extending an orthonormal basis for Span$\{\mathbf{q}_1,\ldots,\mathbf{q}_n\}$ to an orthonormal basis for $\mathbb{R}^m$. Here is one algorithm. Let $\mathbf{e}_1,\ldots,\mathbf{e}_m$ be the standard basis for $\mathbb{R}^m$. Let $\mathbf{f}_1$ be the first vector in this basis that is *not* in W_n = Span$\{\mathbf{q}_1,\ldots,\mathbf{q}_n\}$, and let $\mathbf{u}_1 = \mathbf{f}_1 - \text{proj}_{W_n}\mathbf{f}_1$. Then $\{\mathbf{q}_1,\ldots,\mathbf{q}_n,\mathbf{u}_1\}$ is an orthogonal basis for W_{n+1} = Span$\{\mathbf{q}_1,\ldots,\mathbf{q}_n,\mathbf{u}_1\}$. Let $\mathbf{f}_2$ be the first vector in $\{\mathbf{e}_1,\ldots,\mathbf{e}_m\}$ that is not in W_{n+1}. (Of course $\mathbf{f}_2$ occurs later than $\mathbf{f}_1$ in the list $\mathbf{e}_1,\ldots,\mathbf{e}_m$.) Form $\mathbf{u}_2 = \mathbf{f}_2 - \text{proj}_{W_{n+1}}\mathbf{f}_2$ and W_{n+2} = Span$\{\mathbf{q}_1,\ldots,\mathbf{q}_n,\mathbf{u}_1,\mathbf{u}_2\}$. This process will continue until $m - n$ vectors have been added to the original n vectors. Normalizing the new vectors produces an orthonormal basis for $\mathbb{R}^m$.

MATLAB If A has only two columns, then the Gram-Schmidt process can be implemented with the commands

```
v1 = A(:,1)
v2 = A(:,2) - (A(:,2)'*v1)/(v1'*v1)*v1
```

If A has three columns, add the command

```
v3 = A(:,3) - (A(:,3)'*v1)/(v1'*v1)*v1 - (A(:,3)'*v2)/(v2'*v2)*v2
```

You should use these commands for awhile, to learn the general procedure. After that, you can use the Toolbox command **proj(x,V)** , which computes the projection of a vector **x** onto the subspace spanned by the columns of a matrix (or vector) V. For example,

```
v2 = A(:,2) - proj(A(:,2),v1)            V = v1
v3 = A(:,3) - proj(A(:,3),[v1  v2])      V = [v1  v2]
```

The columns of V in **proj(x,V)** need not be orthogonal for the command to work, but if they are, the entries in **proj(x,V)** will usually agree with those computed via Theorem 10 in Section 7.3, to twelve or more decimal places. Enter **help proj** to learn more about **proj**.

To check your work or to save time, enter **Q = gs(A)** , which uses the Gram-Schmidt process to construct the columns of Q. See **help gs** .

Although you should construct the QR factorization of a matrix using the approach in the text, you might like to see what MATLAB does. The command **[Q1 R1] = qr(A)** creates a modified QR factorization of an $m \times n$ matrix A as described in Exercise 19 in the text. If rank $A = r$, then the first r rows of R_1 are nonzero and the first r columns of Q_1 form an orthonormal basis for Col A.

7.5 LEAST-SQUARES PROBLEMS

The basic geometric principles in this section provide the foundation for all the applications in Sections 7.6—7.8.

KEY IDEAS

You need to know everything on page 367—the definition, the figures, and the discussion. Be careful to distinguish between $\hat{\mathbf{x}}$ and $\hat{\mathbf{b}}$. The least-squares solution is $\hat{\mathbf{x}}$, not $\hat{\mathbf{b}}$. See Fig. 2.

The vector $\hat{\mathbf{b}}$ is unique since it is the closest point in Col A to $\mathbf{b}$. The least-squares solution $\hat{\mathbf{x}}$ is unique if and only if A has linearly independent columns (see Theorems 13 and 14).

Warning: A common mistake is to think that $\hat{\mathbf{x}}$ itself somehow has the least-squares norm or is the closest point to $\mathbf{b}$. Look at Fig. 2 again. The vector closest to $\mathbf{b}$ is $A\hat{\mathbf{x}}$, not $\hat{\mathbf{x}}$.

Theorem 13 provides the main tool for finding least-squares solutions. One way to remember the normal equations is to observe that they are the same as $A\mathbf{x} = \mathbf{b}$, with A^T left-multiplied on each side of the equation. It is completely wrong, however, to try to *derive* the normal equations from $A\mathbf{x} = \mathbf{b}$ by multiplying by A^T. If the equation $A\mathbf{x} = \mathbf{b}$ has no solution, then the equation itself is a false statement about every $\mathbf{x}$. Matrix algebra on such a false statement is meaningless.

The Orthogonal Decomposition Theorem applies in the proof of Theorem 13 even though an orthogonal basis for Col A is not specified, because we have shown that (*i*) every subspace has an orthogonal basis and (*ii*) the orthogonal projection does not depend on the choice of basis.

SOLUTIONS TO EXERCISES

1. $A = \begin{bmatrix} -1 & 2 \\ 2 & -3 \\ -1 & 3 \end{bmatrix}$, $\mathbf{b} = \begin{bmatrix} 4 \\ 1 \\ 2 \end{bmatrix}$,

$$A^TA = \begin{bmatrix} -1 & 2 & -1 \\ 2 & -3 & 3 \end{bmatrix}\begin{bmatrix} -1 & 2 \\ 2 & -3 \\ -1 & 3 \end{bmatrix} = \begin{bmatrix} 6 & -11 \\ -11 & 22 \end{bmatrix}, \quad A^T\mathbf{b} = \begin{bmatrix} -1 & 2 & -1 \\ 2 & -3 & 3 \end{bmatrix}\begin{bmatrix} 4 \\ 1 \\ 2 \end{bmatrix} = \begin{bmatrix} -4 \\ 11 \end{bmatrix}$$

a. The normal equations: $\begin{bmatrix} 6 & -11 \\ -11 & 22 \end{bmatrix}\begin{bmatrix} x_1 \\ x_2 \end{bmatrix} = \begin{bmatrix} -4 \\ 11 \end{bmatrix}$

b. Since A^TA is only 2×2, $(A^TA)^{-1}$ is easy to compute, and

$$\hat{\mathbf{x}} = \begin{bmatrix} 6 & -11 \\ -11 & 22 \end{bmatrix}^{-1} \begin{bmatrix} -4 \\ 11 \end{bmatrix} = \frac{1}{11}\begin{bmatrix} 22 & 11 \\ 11 & 6 \end{bmatrix}\begin{bmatrix} -4 \\ 11 \end{bmatrix} = \frac{1}{11}\begin{bmatrix} 33 \\ 22 \end{bmatrix} = \begin{bmatrix} 3 \\ 2 \end{bmatrix}$$

Warning: It is important to distinguish between the normal equations $A^TA\hat{\mathbf{x}} = A^T\mathbf{b}$ and the formula $\hat{\mathbf{x}} = (A^TA)^{-1}A^T\mathbf{b}$. Both equations describe $\hat{\mathbf{x}}$ (implicitly or explicitly), but the formula for $\hat{\mathbf{x}}$ holds only when A has linearly independent columns. Note that the expression $(A^TA)^{-1}A^T$ *cannot* be simplified when A is not invertible.

7. $A = \begin{bmatrix} 1 & -2 \\ -1 & 2 \\ 0 & 3 \\ 2 & 5 \end{bmatrix}$, $\mathbf{b} = \begin{bmatrix} 3 \\ 1 \\ -4 \\ 2 \end{bmatrix}$. The value of $\hat{\mathbf{x}}$ was found in Exercise 3, but here are the calculations:

$$A^TA = \begin{bmatrix} 1 & -1 & 0 & 2 \\ -2 & 2 & 3 & 5 \end{bmatrix}\begin{bmatrix} 1 & -2 \\ -1 & 2 \\ 0 & 3 \\ 2 & 5 \end{bmatrix} = \begin{bmatrix} 6 & 6 \\ 6 & 42 \end{bmatrix}$$

$$A^T\mathbf{b} = \begin{bmatrix} 1 & -1 & 0 & 2 \\ -2 & 2 & 3 & 5 \end{bmatrix}\begin{bmatrix} 3 \\ 1 \\ -4 \\ 2 \end{bmatrix} = \begin{bmatrix} 6 \\ -6 \end{bmatrix}$$

The normal equations: $\begin{bmatrix} 6 & 6 \\ 6 & 42 \end{bmatrix}\begin{bmatrix} x_1 \\ x_2 \end{bmatrix} = \begin{bmatrix} 6 \\ -6 \end{bmatrix}$

The particular numbers in A^TA suggest that the normal equations might be solved easily via row operations:

$$\begin{bmatrix} 6 & 6 & 6 \\ 6 & 42 & -6 \end{bmatrix} \sim \begin{bmatrix} 6 & 6 & 6 \\ 0 & 36 & -12 \end{bmatrix} \sim \begin{bmatrix} 1 & 1 & 1 \\ 0 & 1 & -1/3 \end{bmatrix} \sim \begin{bmatrix} 1 & 0 & 4/3 \\ 0 & 1 & -1/3 \end{bmatrix}$$

Thus $\hat{\mathbf{x}} = \begin{bmatrix} 4/3 \\ -1/3 \end{bmatrix}$. The least-squares error is $\|A\hat{\mathbf{x}} - \mathbf{b}\|$, so compute

$$A\hat{\mathbf{x}} - \mathbf{b} = \begin{bmatrix} 1 & -2 \\ -1 & 2 \\ 0 & 3 \\ 2 & 5 \end{bmatrix}\begin{bmatrix} 4/3 \\ -1/3 \end{bmatrix} - \begin{bmatrix} 3 \\ 1 \\ -4 \\ 2 \end{bmatrix} = \begin{bmatrix} 2 \\ -2 \\ -1 \\ 1 \end{bmatrix} - \begin{bmatrix} 3 \\ 1 \\ -4 \\ 2 \end{bmatrix} = \begin{bmatrix} -1 \\ -3 \\ 3 \\ -1 \end{bmatrix}$$

$$\|A\hat{\mathbf{x}} - \mathbf{b}\|^2 = 1 + 9 + 9 + 1 = 20, \text{ and } \|A\hat{\mathbf{x}} - \mathbf{b}\| = \sqrt{20} = 2\sqrt{5}$$

Study Tip: A good way to check your work is to verify that $A\hat{\mathbf{x}} - \mathbf{b}$ is orthogonal to each column of A.

13. $A\mathbf{u} = \begin{bmatrix} 3 & 4 \\ -2 & 1 \\ 3 & 4 \end{bmatrix}\begin{bmatrix} 5 \\ -1 \end{bmatrix} = \begin{bmatrix} 11 \\ -11 \\ 11 \end{bmatrix}$, $\mathbf{b} - A\mathbf{u} = \begin{bmatrix} 11 \\ -9 \\ 5 \end{bmatrix} - \begin{bmatrix} 11 \\ -11 \\ 11 \end{bmatrix} = \begin{bmatrix} 0 \\ 2 \\ -6 \end{bmatrix}$, $\|\mathbf{b} - A\mathbf{u}\| = \sqrt{40}$

$A\mathbf{v} = \begin{bmatrix} 3 & 4 \\ -2 & 1 \\ 3 & 4 \end{bmatrix}\begin{bmatrix} 5 \\ -2 \end{bmatrix} = \begin{bmatrix} 7 \\ -12 \\ 7 \end{bmatrix}$, $\mathbf{b} - A\mathbf{v} = \begin{bmatrix} 11 \\ -9 \\ 5 \end{bmatrix} - \begin{bmatrix} 7 \\ -12 \\ 7 \end{bmatrix} = \begin{bmatrix} 4 \\ 3 \\ -2 \end{bmatrix}$, $\|\mathbf{b} - A\mathbf{u}\| = \sqrt{29}$

Obviously, $A\mathbf{u}$ is not the closest point of Col A to $\mathbf{b}$, because $A\mathbf{v}$ is closer. Hence $\mathbf{u}$ is *not* the least-squares solution of $A\mathbf{x} = \mathbf{b}$.

19. a. If A has linearly independent columns, then the equation $A\mathbf{x} = \mathbf{0}$ has only the trivial solution. By Exercise 17, $A^TA\mathbf{x} = \mathbf{0}$ also has only the trivial solution. Since A^TA is *square*, it must be invertible, by the Invertible Matrix Theorem.

b. Since the n linearly independent columns of A belong to $\mathbb{R}^m$, m could not be less than n.

c. The n linearly independent columns of A form a basis for Col A, so the rank of A is n.

MATLAB For Exercises 15 and 16, see the Numerical Note on page 372 in the text. The MATLAB "backslash" command **R\(Q'*b)** will solve $R\hat{\mathbf{x}} = Q^T\mathbf{b}$ by row reduction. When A is nonsquare with linearly independent columns, this procedure of first forming Q and R and then solving $R\hat{\mathbf{x}} = Q^T\mathbf{b}$ is what MATLAB itself does (internally) when the MATLAB command **A\b** is used to solve $A\mathbf{x} = \mathbf{b}$. Compute **A\b** for the data in Exercises 15 and 16, and compare the results with your calculations involving Q and R.

For Exercise 24, the command **A = [A1;A2]** creates a (partitioned) matrix whose top block is A1 and bottom block is A2. This command works as long as A1 and A2 have the same number of columns.

7.6 APPLICATIONS TO LINEAR MODELS

This section will be a valuable reference for any person who works with data that requires statistical analysis. Many graduate fields require such work, often in connection with doctoral research. Even most undergraduates will take a course where least-squares lines are used.

KEY IDEA

Linear algebra unifies the study of many problems in statistics and data analysis. All the examples in this section, from ordinary linear regression (using a least-squares line) to multiple regression, concern just one idea: find a least-squares solution of $X\boldsymbol{\beta} = \mathbf{y}$. Only the design matrix X varies. The exercises help you practice choosing X. The least-squares solution $\hat{\boldsymbol{\beta}}$ always satisfies the normal equations $X^TX\hat{\boldsymbol{\beta}} = X^T\mathbf{y}$.

STUDY NOTES

Don't confuse the least-squares line in Fig. 1 with the lines and planes in Section 7.5 onto which we projected various vectors $\mathbf{b}$. The line is nothing more than a special case of the curves in Figures 2-5. In each case, the "linearity" of the model lies not in the curve, but rather in the fact that the unknown parameters (or *weights*) β_0, $\beta_1, \ldots$ occur linearly in the formula for the curve, just as the variables x_1, $x_2, \ldots$ occur in an ordinary linear equation.

Any 4×4 submatrix of the design matrix in Example 3 is called a Vandermonde matrix. Using Exercise 11 on page 158, one can show that if at least four of the values $x_1, \ldots, x_n$ are distinct, then the least-squares solution $\hat{\boldsymbol{\beta}}$ will be unique, by Theorem 14 in Section 7.5.

FURTHER READING

An important generalization of the discussion here is to *multivariate* analysis, which involves several $\mathbf{y}$ vectors rather than just one. In this case the basic equation is $XB = Y$, where each column of Y is a data set for one dependent variable, and each column of B is a set of parameters to be determined. That is, $X[\boldsymbol{\beta}_1 \ \cdots \ \boldsymbol{\beta}_p] = [\mathbf{y}_1 \ \cdots \ \mathbf{y}_p]$. For more information, see the classic text by T. W. Anderson, *An Introduction to Multivariate Statistical Analysis*, John Wiley & Sons, New York, 1984 (and 1958). The preface of the text says, "A knowledge of matrix algebra is a prerequisite [for understanding the text]." Most modern multivariate statistics texts rely heavily on matrix notation and matrix algebra.

SOLUTIONS TO EXERCISES

1. Place the x-coordinates of the data in the second column of X and the y-coordinates in the vector $\mathbf{y}$. So $X = \begin{bmatrix} 1 & 0 \\ 1 & 1 \\ 1 & 2 \\ 1 & 3 \end{bmatrix}$ and $\mathbf{y} = \begin{bmatrix} 1 \\ 1 \\ 2 \\ 2 \end{bmatrix}$. Compute

$$\underbrace{\begin{bmatrix} 1 & 1 & 1 & 1 \\ 0 & 1 & 2 & 3 \end{bmatrix}}_{X^T}\underbrace{\begin{bmatrix} 1 & 0 \\ 1 & 1 \\ 1 & 2 \\ 1 & 3 \end{bmatrix}}_{X} = \begin{bmatrix} 4 & 6 \\ 6 & 14 \end{bmatrix}, \quad \underbrace{\begin{bmatrix} 1 & 1 & 1 & 1 \\ 0 & 1 & 2 & 3 \end{bmatrix}}_{X^T}\underbrace{\begin{bmatrix} 1 \\ 1 \\ 2 \\ 2 \end{bmatrix}}_{\mathbf{y}} = \begin{bmatrix} 6 \\ 11 \end{bmatrix}$$

The matrix normal equation and its solution are:

$$\begin{bmatrix} 4 & 6 \\ 6 & 14 \end{bmatrix}\begin{bmatrix} \beta_0 \\ \beta_1 \end{bmatrix} = \begin{bmatrix} 6 \\ 11 \end{bmatrix}$$

$$\begin{bmatrix} \beta_0 \\ \beta_1 \end{bmatrix} = \begin{bmatrix} 4 & 6 \\ 6 & 14 \end{bmatrix}^{-1}\begin{bmatrix} 6 \\ 11 \end{bmatrix} = \frac{1}{20}\begin{bmatrix} 14 & -6 \\ -6 & 4 \end{bmatrix}\begin{bmatrix} 6 \\ 11 \end{bmatrix} = \frac{1}{20}\begin{bmatrix} 18 \\ 8 \end{bmatrix} = \begin{bmatrix} .9 \\ .4 \end{bmatrix}$$

The least-squares line, $y = \beta_0 + \beta_1 x$, is $y = .9 + .4x$

7. $\mathbf{y} = X\boldsymbol{\beta} + \boldsymbol{\varepsilon}$, where $\mathbf{y} = \begin{bmatrix} 1.8 \\ 2.7 \\ 3.4 \\ 3.8 \\ 3.9 \end{bmatrix}$, $X = \begin{bmatrix} 1 & 1 \\ 2 & 4 \\ 3 & 9 \\ 4 & 16 \\ 5 & 25 \end{bmatrix}$, $\boldsymbol{\beta} = \begin{bmatrix} \beta_1 \\ \beta_2 \end{bmatrix}$, $\boldsymbol{\varepsilon} = \begin{bmatrix} \varepsilon_1 \\ \varepsilon_2 \\ \varepsilon_3 \\ \varepsilon_4 \\ \varepsilon_5 \end{bmatrix}$.

The MATLAB "backslash" command $\boldsymbol{\beta}$ = **(X'*X)\\(X'*y)** solves the equation $(X^TX)\boldsymbol{\beta} = X^T\mathbf{y}$. For the matrices in this problem, MATLAB produces

$$\begin{bmatrix} \beta_1 \\ \beta_2 \end{bmatrix} = \begin{bmatrix} 1.76 \\ -.20 \end{bmatrix} \quad \text{(to two decimal places)}$$

The desired least-squares equation is $y = 1.76x - .20x^2$.

13. Let $\mathbf{1} = \begin{bmatrix} 1 \\ 1 \\ 1 \\ 1 \end{bmatrix}$ and $\mathbf{x} = \begin{bmatrix} -1 \\ 3 \\ 6 \\ 12 \end{bmatrix}$. Then $\mathbf{v} = \mathbf{x} - \dfrac{\mathbf{x}^T\mathbf{1}}{\mathbf{1}^T\mathbf{1}}\mathbf{1} = \begin{bmatrix} -1 \\ 3 \\ 6 \\ 12 \end{bmatrix} - \dfrac{20}{4}\begin{bmatrix} 1 \\ 1 \\ 1 \\ 1 \end{bmatrix} = \begin{bmatrix} -6 \\ -2 \\ 1 \\ 7 \end{bmatrix}$.

The number 20/4 is the average $\bar{x}$ of the entries in $\mathbf{x}$. So the entries in $\mathbf{v}$ are formed by subtracting $\bar{x}$ from the corresponding entries in $\mathbf{x}$.

15. From Eq. (1),

$$X^TX = \begin{bmatrix} 1 & \cdots & 1 \\ x_1 & \cdots & x_n \end{bmatrix}\begin{bmatrix} 1 & x_1 \\ \vdots & \vdots \\ 1 & x_n \end{bmatrix} = \begin{bmatrix} n & \Sigma x \\ \Sigma x & \Sigma x^2 \end{bmatrix}$$

$$X^T\mathbf{y} = \begin{bmatrix} 1 & \cdots & 1 \\ x_1 & \cdots & x_n \end{bmatrix}\begin{bmatrix} y_1 \\ \vdots \\ y_n \end{bmatrix} = \begin{bmatrix} \Sigma y \\ \Sigma xy \end{bmatrix}$$

The equations (6) in the text follow immediately from the usual matrix normal equation $X^TX\boldsymbol{\beta} = X^T\mathbf{y}$.

18. *Note*: The formulas you should derive are

$$\hat{\beta}_0 = \frac{\Sigma x^2\Sigma y - \Sigma x\Sigma xy}{n\Sigma x^2 - (\Sigma x)^2}, \qquad \hat{\beta}_1 = \frac{n\Sigma xy - \Sigma x\Sigma y}{n\Sigma x^2 - (\Sigma x)^2}$$

Some statistics texts present other equivalent formulas for $\hat{\beta}_0$ and $\hat{\beta}_1$.

19. The equation to be proved is $\|\mathbf{y}\|^2 = \|X\hat{\boldsymbol{\beta}}\|^2 + \|\mathbf{y} - X\hat{\boldsymbol{\beta}}\|^2$. This follows from the Pythagorean Theorem (in Section 7.1) and the figure below.

Appendix: The Geometry of a Linear Model

The column space of the design matrix X is sometimes called the **design subspace**. If $\hat{\boldsymbol{\beta}}$ is the least-squares solution of $\mathbf{y} = X\boldsymbol{\beta}$, then the residual vector $\boldsymbol{\varepsilon} = \mathbf{y} - X\hat{\boldsymbol{\beta}}$ is orthogonal to the design subspace, and the equation $\mathbf{y} = X\hat{\boldsymbol{\beta}} + \boldsymbol{\varepsilon}$ is an orthogonal decomposition of the observed $\mathbf{y}$ into the sum of the least-squares predicted $\hat{\mathbf{y}}$ and the residual vector $\boldsymbol{\varepsilon}$.

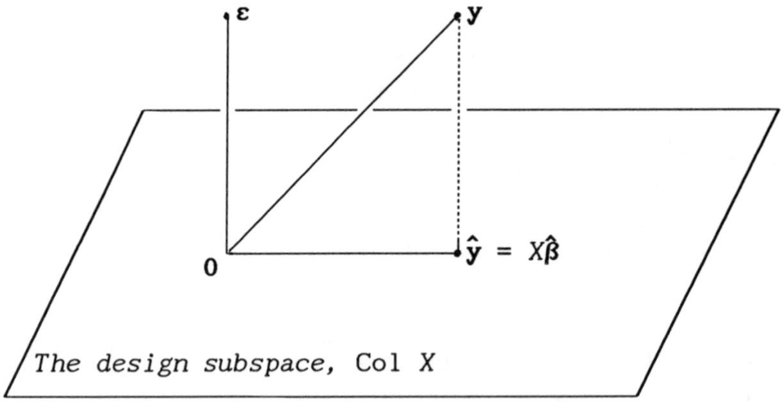

7.7 INNER PRODUCT SPACES

Three examples of inner product spaces are described here, in Examples 1, 2, and 7. Corresponding applications appear in the next section. Material in Sections 7.7 and 7.8 will be useful for many careers, particularly science, engineering, and mathematics. If your course does not cover this now, the Study Guide can help you learn it later on your own.

KEY IDEAS

The concepts of length and orthogonality in $\mathbb{R}^n$ have analogues in a number of other vector spaces. The definition of an inner product identifies the basic properties needed for a theory that parallels the familiar theory for $\mathbb{R}^n$. Two useful facts, the Cauchy-Schwarz inequality and the triangle inequality, were not developed earlier, but they are important for applications both in $\mathbb{R}^n$ and in other inner product spaces. Every mathematics major will need to know these facts in other undergraduate courses.

The inner product in Example 1 is used in Section 7.8 to describe weighted least-squares problems. The inner product in Examples 2-6 provides a more sophisticated approach to the least-squares curve fitting discussed in Section 7.6. See the "trend analysis" in Section 7.8.

Be sure to read and reread the paragraph at the bottom of page 387. The idea of "best approximation" to a function is of fundamental importance in mathematics. The most common applications of best approximation (such as Fourier series, introduced in Section 7.8) involve the inner product in Example 7.

SOLUTIONS TO EXERCISES

1. The inner product is $\langle \mathbf{x}, \mathbf{y} \rangle = 4x_1y_1 + 5x_2y_2$. Let $\mathbf{x} = (1,1)$, $\mathbf{y} = (5,-1)$.

a. $\|\mathbf{x}\|^2 = 4\cdot1\cdot1 + 5\cdot1\cdot1 = 9$, $\|\mathbf{x}\| = 3$

$\|\mathbf{y}\|^2 = 4\cdot5\cdot5 + 5(-1)(-1) = 105$, $\|\mathbf{y}\| = \sqrt{105}$

b. A vector $\mathbf{z} = (z_1, z_2)$ is orthogonal to $\mathbf{y}$ if and only if $\langle \mathbf{z}, \mathbf{y} \rangle = 0$, that is,

$$4\cdot z_1\cdot 5 + 5\cdot z_2\cdot(-1) = 0, \quad 20z_1 - 5z_2 = 0, \quad \text{and} \quad z_2 = 4z_1$$

Thus (z_1, z_2) is orthogonal to $\mathbf{y}$ if and only if $z_2 = 4z_1$.

7. Given $p(t) = 4 + t$ and $q(t) = 5 - 4t^2$. The orthogonal projection $\hat{q}$ of q onto the subspace spanned by p is $[\langle q,p\rangle/\langle p,p\rangle]p$. The notation of Example 5 organizes the calculations nicely:

	p	q	
Polynomial:			
Vector of values:	$\begin{bmatrix}3\\4\\5\end{bmatrix}$,	$\begin{bmatrix}1\\5\\1\end{bmatrix}$	← value at -1 ← value at 0 ← value at 1

The inner product $\langle q,p\rangle$ equals the (standard) inner product of the two corresponding vectors in $\mathbb{R}^3$: $\langle q,p\rangle = 3\cdot 1 + 4\cdot 5 + 5\cdot 1 = 28$. Similarly, $\langle p,p\rangle = 3^2 + 4^2 + 5^2 = 50$. Thus

$$\hat{q}(t) = \frac{28}{50}(4 + t) = \frac{56}{25} + \frac{14}{25}t$$

13. Suppose A is invertible and $\langle \mathbf{u},\mathbf{v}\rangle = (A\mathbf{u})\cdot(A\mathbf{v})$, for $\mathbf{u},\mathbf{v}$ in $\mathbb{R}^n$. We check each axiom in the definition on page 384.

i. $\langle \mathbf{u},\mathbf{v}\rangle = (A\mathbf{u})\cdot(A\mathbf{v}) = (A\mathbf{v})\cdot(A\mathbf{u})$ Property of dot product
$= \langle \mathbf{v},\mathbf{u}\rangle$

ii. $\langle \mathbf{u}+\mathbf{v},\mathbf{w}\rangle = [A(\mathbf{u}+\mathbf{v})]\cdot(A\mathbf{w}) = [A\mathbf{u}+A\mathbf{v}]\cdot(A\mathbf{w})$ Matrix multiplication
$= (A\mathbf{u})\cdot(A\mathbf{w}) + (A\mathbf{v})\cdot(A\mathbf{w})$ Property of dot product
$= \langle \mathbf{u},\mathbf{w}\rangle + \langle \mathbf{v},\mathbf{w}\rangle$

iii. $\langle c\mathbf{u},\mathbf{v}\rangle = [A(c\mathbf{u})]\cdot(A\mathbf{v}) = [c(A\mathbf{u})]\cdot(A\mathbf{v})$ Matrix multiplication
$= c(A\mathbf{u})\cdot(A\mathbf{v})$ Property of dot product
$= c\langle \mathbf{u},\mathbf{v}\rangle$

iv. $\langle \mathbf{u},\mathbf{u}\rangle = (A\mathbf{u})\cdot(A\mathbf{u}) = \|A\mathbf{u}\|^2 \geq 0$, and this quantity is zero if and only if the vector $A\mathbf{u}$ is $\mathbf{0}$. But $A\mathbf{u} = \mathbf{0}$ if and only if $\mathbf{u} = \mathbf{0}$, because A is invertible.

Another method for verifying the axioms is to use properties of the transpose operation. The calculations are similar. However, for (i), you need to use the fact that the transpose of a scalar (which is a 1×1 matrix) is the scalar itself: $\langle \mathbf{u},\mathbf{v}\rangle = \langle \mathbf{u},\mathbf{v}\rangle^T = [(A\mathbf{u})^T(A\mathbf{v})]^T = (A\mathbf{v})^T(A\mathbf{u})^{TT} = (A\mathbf{v})^T(A\mathbf{u}) = \langle \mathbf{v},\mathbf{u}\rangle$.

17. $\|\mathbf{u}+\mathbf{v}\|^2 = \langle \mathbf{u}+\mathbf{v},\mathbf{u}+\mathbf{v}\rangle = \langle \mathbf{u},\mathbf{u}+\mathbf{v}\rangle + \langle \mathbf{v},\mathbf{u}+\mathbf{v}\rangle$ Axiom ii
$= \langle \mathbf{u}+\mathbf{v},\mathbf{u}\rangle + \langle \mathbf{u}+\mathbf{v},\mathbf{v}\rangle$ Axiom i
$= \langle \mathbf{u},\mathbf{u}\rangle + \langle \mathbf{v},\mathbf{u}\rangle + \langle \mathbf{u},\mathbf{v}\rangle + \langle \mathbf{v},\mathbf{v}\rangle$ Axiom ii
$= \langle \mathbf{u},\mathbf{u}\rangle + 2\langle \mathbf{u},\mathbf{v}\rangle + \langle \mathbf{v},\mathbf{v}\rangle$ Axiom i

Next, replace **v** by −**v** and use the fact that −**v** = (−1)**v**.

$$\|\mathbf{u} - \mathbf{v}\|^2 = \langle \mathbf{u},\mathbf{u}\rangle + 2\langle \mathbf{u},-\mathbf{v}\rangle + \langle -\mathbf{v},-\mathbf{v}\rangle \qquad \text{Replacing } \mathbf{v} \text{ above by } -\mathbf{v}.$$
$$= \langle \mathbf{u},\mathbf{u}\rangle - 2\langle \mathbf{u},\mathbf{v}\rangle + (-1)^2\langle \mathbf{v},\mathbf{v}\rangle \qquad \text{Axiom iii and Exercise 15}$$

Subtracting, $\|\mathbf{u} + \mathbf{v}\|^2 - \|\mathbf{u} - \mathbf{v}\|^2 = 2\langle \mathbf{u},\mathbf{v}\rangle - (-2\langle \mathbf{u},\mathbf{v}\rangle) = 4\langle \mathbf{u},\mathbf{v}\rangle$. Division by 4 gives the desired identity.

19. The full solution is in the text.

25. In the space $C[-1,1]$ with the integral inner product, the polynomials t and 1 are orthogonal, because

$$\langle t,1\rangle = \int_{-1}^{1} t\cdot 1\,dt = \frac{1}{2}t^2\Big|_{-1}^{1} = \frac{1}{2}(1)^2 - \frac{1}{2}(-1)^2 = 0$$

So 1 and t can be in an orthogonal basis for Span$\{1, t, t^2\}$. Next, compute $\text{proj}_W t^2$, the orthogonal projection of the vector t^2 onto the subspace W spanned by 1 and t.

$$\langle t^2,1\rangle = \int_{-1}^{1} t^2\cdot 1\,dt = \frac{1}{3}t^3\Big|_{-1}^{1} = \frac{1}{3}(1)^3 - \frac{1}{3}(-1)^3 = \frac{2}{3}$$

$$\langle 1,1\rangle = \int_{-1}^{1} 1\cdot 1\,dt = t\Big|_{-1}^{1} = 1 - (-1) = 2$$

$$\langle t^2,t\rangle = \int_{-1}^{1} t^2\cdot t\,dt = \frac{1}{4}t^4\Big|_{-1}^{1} = \frac{1}{4}(1)^4 - \frac{1}{4}(-1)^4 = 0$$

There is no need to compute $\langle t,t\rangle$, because t^2 is orthogonal to t. Thus

$$\text{proj}_W t^2 = \frac{\langle t^2,1\rangle}{\langle 1,1\rangle}1 + \frac{\langle t^2,t\rangle}{\langle t,t\rangle}t = \frac{2/3}{2}1 + 0 = \frac{1}{3}$$

A polynomial orthogonal to W is $t^2 - \text{proj}_W t^2 = t^2 - \frac{1}{3}$. Another choice is this polynomial scaled by 3, namely, $3t^2 - 1$. Thus, the polynomials, 1, t^2, and $3t^2 - 1$ form an orthogonal basis for Span$\{1, t, t^2\}$.

Can you find the next Legendre polynomial, a cubic polynomial that is orthogonal to the first three Legendre polynomials?

7.8 APPLICATIONS OF INNER PRODUCT SPACES

Of the three applications in this section, the discussion of Fourier series is by far the most important. Such series have great practical value, particularly in mathematics, engineering and the physical sciences. Calculations with Fourier series are simple because sine and cosine functions are orthogonal. This fact is often overlooked in undergraduate courses that do not assume a linear algebra background.

KEY IDEAS

The text gives the normal equation for the weighted least-squares solution of $A\mathbf{x} = \mathbf{y}$. When applied to a least-squares line problem, the most common situation, the normal equation is usually written as

$$(WX)^T WX\hat{\boldsymbol{\beta}} = (WX)^T\mathbf{y}$$

where W is the (diagonal) weighting matrix, X is the design matrix, $\hat{\boldsymbol{\beta}}$ is the least-squares parameter vector, and $\mathbf{y}$ is the observation vector.

Trend analysis is really a least-squares regression problem of the type described in Section 7.6, with data points $(x_1, y_1), \ldots, (x_n, y_n)$ fitted by a curve of the form

$$y = \beta_0 f_0(x) + \beta_1 f_1(x) + \cdots + \beta_k f_k(x)$$

where the functions $f_0, \ldots, f_k$ are polynomials that are orthogonal with respect to an inner product on $\mathbb{P}_{n-1}$ defined by

$$\langle p, q\rangle = p(x_1)q(x_1) + \cdots + p(x_n)q(x_n)$$

Usually, $x_1, \ldots, x_n$ are arranged to be evenly spaced and sum to zero, and the functions $f_1, \ldots, f_k$ are of degree 3 or 4 or less.

In $C[0, 2\pi]$ with the integral inner product, the set

$$\{1, \cos t, \cos 2t, \ldots, \cos nt, \sin t, \sin 2t, \ldots, \sin nt\} \qquad (*)$$

is orthogonal. The nth order Fourier approximation to some f in $C[0, 2\pi]$ is simply the orthogonal projection of f onto the subspace W of trigonometric polynomials spanned by the functions in (*). The Fourier coefficients of f are the weights in the usual formula for the orthogonal projection of f onto W.

If an application involves an interval $[0, T]$ instead of $[0, 2\pi]$, then the inner product requires an integral over $[0, T]$, and the appropriate orthogonal set is obtained by replacing t in each function in (*) with $2\pi t/T$.

SOLUTIONS TO EXERCISES

1. For the data (-2,0), (-1,0), (0,2), (1,4), (2,4), construct

$$X = \begin{bmatrix} 1 & -2 \\ 1 & -1 \\ 1 & 0 \\ 1 & 1 \\ 1 & 2 \end{bmatrix} \text{ Design matrix} \qquad \boldsymbol{\beta} = \begin{bmatrix} \beta_0 \\ \beta_1 \end{bmatrix} \text{ Parameter vector} \qquad \mathbf{y} = \begin{bmatrix} 0 \\ 0 \\ 2 \\ 4 \\ 4 \end{bmatrix} \text{ Observation vector}$$

Since the first and last data points are about half as reliable as the other points, a suitable weighting matrix is

$$W = \begin{bmatrix} 1 & 0 & 0 & 0 & 0 \\ 0 & 2 & 0 & 0 & 0 \\ 0 & 0 & 2 & 0 & 0 \\ 0 & 0 & 0 & 2 & 0 \\ 0 & 0 & 0 & 0 & 1 \end{bmatrix}, \text{ so } WX = \begin{bmatrix} 1 & -2 \\ 2 & -2 \\ 2 & 0 \\ 2 & 2 \\ 1 & 2 \end{bmatrix} \text{ and } W\mathbf{y} = \begin{bmatrix} 0 \\ 0 \\ 4 \\ 8 \\ 4 \end{bmatrix}$$

The remaining calculations are the same as in ordinary least-squares, except that the *weighted* design matrix WX and the *weighted* observation vector $W\mathbf{y}$ appear in place of X and $\mathbf{y}$.

$$(WX)^T(WX) = \begin{bmatrix} 1 & 2 & 2 & 2 & 1 \\ -2 & -2 & 0 & 2 & 2 \end{bmatrix} \begin{bmatrix} 1 & -2 \\ 2 & -2 \\ 2 & 0 \\ 2 & 2 \\ 1 & 2 \end{bmatrix} = \begin{bmatrix} 14 & 0 \\ 0 & 16 \end{bmatrix}$$

$$(WX)^T(W\mathbf{y}) = \begin{bmatrix} 1 & 2 & 2 & 2 & 1 \\ -2 & -2 & 0 & 2 & 2 \end{bmatrix} \begin{bmatrix} 0 \\ 0 \\ 4 \\ 8 \\ 4 \end{bmatrix} = \begin{bmatrix} 28 \\ 24 \end{bmatrix}$$

The normal equation and solution are

$$\begin{bmatrix} 14 & 0 \\ 0 & 16 \end{bmatrix} \begin{bmatrix} \beta_0 \\ \beta_1 \end{bmatrix} = \begin{bmatrix} 28 \\ 24 \end{bmatrix}, \qquad \begin{bmatrix} \beta_0 \\ \beta_1 \end{bmatrix} = \begin{bmatrix} 1/14 & 0 \\ 0 & 1/16 \end{bmatrix} \begin{bmatrix} 28 \\ 24 \end{bmatrix} = \begin{bmatrix} 2 \\ 3/2 \end{bmatrix}$$

The equation of the least-squares line is $y = 2 + (3/2)x$

7. $$\|\cos kt\|^2 = \int_0^{2\pi} \cos kt \cdot \cos kt \, dt = \int_0^{2\pi} \frac{1 + \cos 2kt}{2} \, dt$$

$$= \left[\frac{1}{2}t + \frac{\sin 2kt}{4k}\right]\Big|_0^{2\pi} = (\tfrac{1}{2}\cdot 2\pi + 0) - 0 = \pi \quad (\text{if } k \neq 0)$$

$$\|\sin kt\|^2 = \int_0^{2\pi} \sin kt \cdot \sin kt\, dt = \int_0^{2\pi} \frac{1 - \cos 2kt}{2}\, dt$$

$$= \left[\frac{1}{2}t - \frac{\sin 2kt}{4k}\right]\Big|_0^{2\pi} = (\tfrac{1}{2}\cdot 2\pi - 0) - 0 = \pi \quad (\text{if } k \neq 0)$$

9. $f(t) = 2\pi - t$. The definite integrals of $t\cos kt$ and $t\sin kt$, shown below, were computed in Example 4. The Fourier coefficients of f are:

$$\frac{a_0}{2} = \frac{1}{2}\cdot\frac{1}{\pi}\int_0^{2\pi} (2\pi - t)\, dt = -\frac{1}{2}\cdot\frac{1}{2\pi}(2\pi - t)^2\Big|_0^{2\pi} = 0 + \frac{1}{4\pi}(2\pi)^2 = \pi$$

and for $k > 0$,

$$a_k = \frac{1}{\pi}\int_0^{2\pi} (2\pi - t)\cos kt\, dt = \frac{1}{\pi}\int_0^{2\pi} 2\pi\cos kt\, dt - \frac{1}{\pi}\int_0^{2\pi} t\cos kt\, dt$$

$$= 0 - 0 = 0$$

$$b_k = \frac{1}{\pi}\int_0^{2\pi} (2\pi - t)\sin kt\, dt = \frac{1}{\pi}\int_0^{2\pi} 2\pi\sin kt\, dt - \frac{1}{\pi}\int_0^{2\pi} t\sin kt\, dt$$

$$= 0 - (-\frac{2}{k}) = \frac{2}{k}$$

The third-order Fourier approximation to f is

$$\pi + 2\sin t + \sin 2t + \frac{2}{3}\sin 3t$$

13. Take f and g in $C[0, 2\pi]$ and let m be a nonnegative integer. Then, the linearity of the inner product shows that

$$\langle (f + g), \cos mt\rangle = \langle f, \cos mt\rangle + \langle g, \cos mt\rangle$$

Dividing each term in this equality by $\langle \cos mt, \cos mt\rangle$, we conclude that the Fourier coefficient a_m of $f + g$ is the sum of the corresponding Fourier coefficients of f and of g. Similarly, the Fourier coefficient b_m of $f + g$ is the sum of the corresponding Fourier coefficients of f and of g.

Appendix: The Linearity of an Orthogonal Projection

The argument for Exercise 13 is a special case of a general principle. In any inner product space, the mapping $\mathbf{y} \mapsto \frac{\langle \mathbf{y}, \mathbf{u} \rangle}{\langle \mathbf{u}, \mathbf{u} \rangle}\mathbf{u}$ is linear, for any nonzero $\mathbf{u}$. To verify this, take any $\mathbf{x}$ and $\mathbf{y}$ in the space and any scalar c. Then

$$\frac{\langle \mathbf{x}+\mathbf{y}, \mathbf{u} \rangle}{\langle \mathbf{u}, \mathbf{u} \rangle}\mathbf{u} = \frac{\langle \mathbf{x}, \mathbf{u} \rangle}{\langle \mathbf{u}, \mathbf{u} \rangle}\mathbf{u} + \frac{\langle \mathbf{y}, \mathbf{u} \rangle}{\langle \mathbf{u}, \mathbf{u} \rangle}\mathbf{u}, \quad \text{and} \quad \frac{\langle c\mathbf{x}, \mathbf{u} \rangle}{\langle \mathbf{u}, \mathbf{u} \rangle}\mathbf{u} = \frac{c\langle \mathbf{x}, \mathbf{u} \rangle}{\langle \mathbf{u}, \mathbf{u} \rangle}\mathbf{u} = c\frac{\langle \mathbf{x}, \mathbf{u} \rangle}{\langle \mathbf{u}, \mathbf{u} \rangle}\mathbf{u}$$

Similarly, if $\mathbf{u}_1, \ldots, \mathbf{u}_p$ are any nonzero vectors, then the mapping

$$\mathbf{y} \mapsto \frac{\langle \mathbf{y}, \mathbf{u}_1 \rangle}{\langle \mathbf{u}_1, \mathbf{u}_1 \rangle}\mathbf{u}_1 + \cdots + \frac{\langle \mathbf{y}, \mathbf{u}_p \rangle}{\langle \mathbf{u}_p, \mathbf{u}_p \rangle}\mathbf{u}_p$$

is a linear transformation. Thus, if $\{\mathbf{u}_1, \ldots, \mathbf{u}_p\}$ is an orthogonal basis for a subspace W, then the mapping $\mathbf{y} \mapsto \text{proj}_W \mathbf{y}$ is a linear transformation. In particular, if W is the vector space of trigonometric polynomials of order at most n, and if f and g are in $C[0, 2\pi]$, then

$$\text{proj}_W(f + g) = \text{proj}_W f + \text{proj}_W g$$

That is, the nth order Fourier approximation to $f + g$ is the sum of the nth order Fourier approximations to f and to g. Can you use the linearity of the mapping $f \mapsto \text{proj}_W f$ and the final result of Example 4, to produce (with practically no work) the answer to Exercise 9? [*Hint*: The nth order Fourier approximation to a constant function is the function itself.]

CHAPTER 7 GLOSSARY CHECKLIST

Check your knowledge by attempting to write definitions of the terms below. Then compare your work with the definitions given in the text's Glossary. Ask your instructor which definitions, if any, might appear on a test.

angle (between nonzero vectors **u** and **v** in $\mathbb{R}^2$ or $\mathbb{R}^3$): The angle ϑ between the .... Related to the scalar product by: $\mathbf{u}\cdot\mathbf{v}$ = ...

best approximation: The closest point ....

Cauchy-Schwarz inequality: ... for all **u**, **v**.

component of y orthogonal to u (for $\mathbf{u} \neq \mathbf{0}$): The vector ....

design matrix: The matrix X in the linear model ..., where the columns of X are determined in some way by ....

distance between u and v: ..., denoted by ....

Fourier approximation (of order n): The closest point in ... to ....

Fourier coefficients: The weights used to make ....

Fourier series: An infinite series that ... in $C[0,2\pi]$, with the inner product given by ....

fundamental subspaces (of A): The ... space of A, and the ... space of ..., with ... commonly called the ... space of A.

general least-squares problem: Given an $m \times n$ matrix A and a vector **b** in $\mathbb{R}^m$, find ... such that ... for all ....

Gram-Schmidt process: An algorithm for producing ....

Householder reflection (matrix): A matrix Q = ..., where ....

inner product: The scalar ..., usually written as $\mathbf{u}\cdot\mathbf{v}$, where **u** and **v** are vectors in $\mathbb{R}^n$ viewed as .... Also called the ... of **u** and **v**. In general, a function on a vector space that assigns to each pair of vectors **u** and **v** a number ..., subject to certain axioms.

inner product space: A ... space on which is defined ....

least-squares error: The ..., when $\hat{\mathbf{x}}$ is a least-squares solution of ....

least-squares line: The line ... that minimizes ... in the equation ....

least-squares solution (of $A\mathbf{x} = \mathbf{b}$): A vector ... such that ....

length (of **v**): The scalar ...; also called the ... of **v**.

linear model (in statistics): Any equation of the form ..., where X and $\mathbf{y}$ are known and β is to be chosen to minimize ....

mean-deviation form (of a vector): A vector whose entries ....

mean square error: The error of ... in an inner product space, where the inner product is defined by ....

multiple regression: A linear model involving ... variables and ....

normal equations: The system of equations represented by ..., whose solution yields all ... solutions of $A\mathbf{x} = \mathbf{b}$. In statistics, a common notation is ....

normalizing (a vector $\mathbf{v}$): The process of creating a ... vector $\mathbf{u}$ that ....

observation vector: The vector ... in the linear model $\mathbf{y} = X\beta + \varepsilon$, where the entries in ... are the observed values of ....

orthogonal basis: A basis that ....

orthogonal complement (of W): The set $W^{\perp}$ of ....

orthogonal matrix: A ... matrix U such that ....

orthogonal projection of y onto u (or onto the line through $\mathbf{u}$ and the origin, for $\mathbf{u} \neq \mathbf{0}$): The vector ... defined by ....

orthogonal projection of y onto W: The unique vector ... such that .... Notation: $\hat{\mathbf{y}} = \mathrm{proj}_W \mathbf{y}$.

orthogonal set: A set S of vectors such that ... for ....

orthogonal to W: Orthogonal to every ....

orthonormal basis: A basis that is ....

orthonormal set: An ... set of ....

parameter vector: The unknown vector ... in the linear model ....

regression coefficients: The coefficients ... in the ....

residual vector: The quantity ... that appears in the general linear model: ..., the difference between ... and the ... values (of y).

QR factorization: A factorization of an $m \times n$ matrix A with linearly independent columns, $A = QR$, where Q is an ... matrix whose ..., and R is an ... matrix.

same direction (as a vector $\mathbf{v}$): A vector that is ....

scale (a vector): Multiply a ... by ....

Schur factorization (of A, for real scalars): A factorization $A = \ldots$ of an $n \times n$ matrix A having ..., where ... is an ... matrix and ... is an ... matrix.

trend analysis: The use of ... to fit data, with the inner product ....

triangle inequality: ....

trigonometric polynomial: A linear combination of ... and ... functions such as ....

unit vector: A vector $\mathbf{v}$ such that ....

weighted least squares: Least-squares problems with a ...inner product such as $\langle \mathbf{x}, \mathbf{y} \rangle = \ldots$.

8

Symmetric Matrices and Quadratic Forms

8.1 DIAGONALIZATION OF SYMMETRIC MATRICES

To prepare for this section, review Section 7.2 and the Diagonalization Theorem of Section 6.3, along with Example 3 and Theorem 6 of Section 6.3.

KEY IDEAS

If a symmetric matrix has distinct eigenvalues, as in Example 2, then the ordinary diagonalization process will produce a matrix P with *orthogonal* columns, because eigenvectors from different eigenspaces are automatically orthogonal. However, the P you need here must have ortho*normal* columns. Forgetting to normalize the columns of P is the main error students make in this section.

The statements in the spectral theorem, together with the general approach used in Example 3, lead to the following outline for orthogonally diagonalizing any symmetric matrix.

Procedure for Orthogonally Diagonalizing a Symmetric Matrix A

1. *Find the characteristic equation of A.*
2. *For each eigenvalue with multiplicity 1, find an eigenvector and normalize it.*
3. *For each eigenvalue with multiplicity $k \geq 2$, find a basis of k vectors for the eigenspace. Then use the Gram-Schmidt process to construct an orthonormal basis.*
4. *The union of the bases for all the eigenspaces is an orthonormal basis for $\mathbb{R}^n$ (if A is $n \times n$). Use these vectors as the columns of an orthogonal matrix P.*

5. *Construct D from the eigenvalues, in an order corresponding to the columns of P. The number of times an eigenvalue appears on the diagonal of D is equal to the dimension of the corresponding eigenspace.*

6. *Finally,* $A = PDP^{-1} = PDP^T$.

SOLUTIONS TO EXERCISES

1. $A = \begin{bmatrix} 3 & 5 \\ 5 & -7 \end{bmatrix} = A^T$, because the (1,2) and (2,1) entries match. The entries on the main diagonal of A can have any values.

7. $P = \begin{bmatrix} .6 & .8 \\ .8 & -.6 \end{bmatrix} = [\mathbf{p}_1 \ \ \mathbf{p}_2]$. To show that P is orthogonal by hand calculations, show that its columns are orthonormal: $\mathbf{p}_1 \cdot \mathbf{p}_2 = .48 - .48 = 0$, $\|\mathbf{p}_1\|^2 = (.6)^2 + (.8)^2 = 1$, and similarly, $\|\mathbf{p}_2\|^2 = 1$. Since P is square, P is an orthogonal matrix.

13. $A = \begin{bmatrix} 3 & 1 \\ 1 & 3 \end{bmatrix}$. Characteristic polynomial: $(3 - \lambda)^2 - 1 = \lambda^2 - 6\lambda + 8 = (\lambda - 4)(\lambda - 2)$. So the eigenvalues are 4 and 2.

<u>For $\lambda = 4$:</u> $[A - 4I \ \ \mathbf{0}] = \begin{bmatrix} -1 & 1 & 0 \\ 1 & -1 & 0 \end{bmatrix} \sim \begin{bmatrix} 1 & -1 & 0 \\ 0 & 0 & 0 \end{bmatrix}$, $\begin{matrix} x_1 = x_2 \\ x_2 \text{ is free} \end{matrix}$

Take $x_2 = 1$ to get a basis for the eigenspace: $\begin{bmatrix} 1 \\ 1 \end{bmatrix}$. Then normalize to get a unit vector: $\mathbf{u}_1 = \begin{bmatrix} 1/\sqrt{2} \\ 1/\sqrt{2} \end{bmatrix}$. (Don't forget this step.)

<u>For $\lambda = 2$:</u> $[A - 2I \ \ \mathbf{0}] = \begin{bmatrix} 1 & 1 & 0 \\ 1 & 1 & 0 \end{bmatrix} \sim \begin{bmatrix} 1 & 1 & 0 \\ 0 & 0 & 0 \end{bmatrix}$, $\begin{matrix} x_1 = -x_2 \\ x_2 \text{ is free} \end{matrix}$

Take $x_2 = 1$ to get a basis for the eigenspace: $\begin{bmatrix} -1 \\ 1 \end{bmatrix}$. Then normalize to get a unit vector: $\mathbf{u}_2 = \begin{bmatrix} -1/\sqrt{2} \\ 1/\sqrt{2} \end{bmatrix}$.

Set $P = [\mathbf{u}_1 \ \ \mathbf{u}_2] = \begin{bmatrix} 1/\sqrt{2} & -1/\sqrt{2} \\ 1/\sqrt{2} & 1/\sqrt{2} \end{bmatrix}$. The corresponding D is $\begin{bmatrix} 4 & 0 \\ 0 & 2 \end{bmatrix}$.

Study Tip: The fact that eigenvectors for distinct eigenvalues are orthogonal gives you a check on your work. After you find $\mathbf{u}_2$ in Exercise 13, verify that $\mathbf{u}_2 \cdot \mathbf{u}_1 = 0$. When $\mathbf{u}_1$ and $\mathbf{u}_2$ are in $\mathbb{R}^2$, you can easily guess what $\mathbf{u}_2$ must be, once you know $\mathbf{u}_1$. If you do this, you should compute $A\mathbf{u}_2$, to make sure that $\mathbf{u}_2$ is indeed an eigenvector.

19. Be sure to *work* this problem before reading the solution. Use Exercises 13-24 to sharpen your skills. They are critical for the rest of the chapter.

Here, $A = \begin{bmatrix} 3 & 2 & 0 \\ 2 & 4 & 2 \\ 0 & 2 & 5 \end{bmatrix}$, and the eigenvalues are given: 7, 4, 1.

For $\lambda = 7$: $[A - 7I \quad 0] = \begin{bmatrix} -4 & 2 & 0 & 0 \\ 2 & -3 & 2 & 0 \\ 0 & 2 & -2 & 0 \end{bmatrix} \sim \begin{bmatrix} 1 & 0 & -1/2 & 0 \\ 0 & 1 & -1 & 0 \\ 0 & 0 & 0 & 0 \end{bmatrix}$

$x_1 = (1/2)x_3$, $x_2 = x_3$, and x_3 is free. Take $x_3 = 2$ to avoid fractions.

A basis for the eigenspace is $\begin{bmatrix} 1 \\ 2 \\ 2 \end{bmatrix}$; a unit vector is $\mathbf{u}_1 = \begin{bmatrix} 1/3 \\ 2/3 \\ 2/3 \end{bmatrix}$.

For $\lambda = 4$: $[A - 4I \quad 0] = \begin{bmatrix} -1 & 2 & 0 & 0 \\ 2 & 0 & 2 & 0 \\ 0 & 2 & 1 & 0 \end{bmatrix} \sim \begin{bmatrix} 1 & 0 & 1 & 0 \\ 0 & 1 & 1/2 & 0 \\ 0 & 0 & 0 & 0 \end{bmatrix}$

$x_1 = -x_3$, $x_2 = -(1/2)x_3$, and x_3 is free. Take $x_3 = 2$ to avoid fractions.

A basis for the eigenspace is $\begin{bmatrix} -2 \\ -1 \\ 2 \end{bmatrix}$; a unit vector is $\mathbf{u}_2 = \begin{bmatrix} -2/3 \\ -1/3 \\ 2/3 \end{bmatrix}$.

(Remember to check mentally that $\mathbf{u}_2 \cdot \mathbf{u}_1 = 0$.)

For $\lambda = 1$: $[A - I \quad 0] = \begin{bmatrix} 2 & 2 & 0 & 0 \\ 2 & 3 & 2 & 0 \\ 0 & 2 & 4 & 0 \end{bmatrix} \sim \begin{bmatrix} 1 & 0 & -2 & 0 \\ 0 & 1 & 2 & 0 \\ 0 & 0 & 0 & 0 \end{bmatrix}$

$x_1 = 2x_3$, $x_2 = -2x_3$, and x_3 is free. Take $x_3 = 1$.

A basis for the eigenspace is $\begin{bmatrix} 2 \\ -2 \\ 1 \end{bmatrix}$; a unit vector is $\mathbf{u}_3 = \begin{bmatrix} 2/3 \\ -2/3 \\ 1/3 \end{bmatrix}$.

(Also, check that $\mathbf{u}_3 \cdot \mathbf{u}_1 = 0$ and $\mathbf{u}_3 \cdot \mathbf{u}_2 = 0$.)

Set $P = [\mathbf{u}_1 \quad \mathbf{u}_2 \quad \mathbf{u}_3] = \begin{bmatrix} 1/3 & -2/3 & 2/3 \\ 2/3 & -1/3 & -2/3 \\ 2/3 & 2/3 & 1/3 \end{bmatrix}$, and $D = \begin{bmatrix} 7 & 0 & 0 \\ 0 & 4 & 0 \\ 0 & 0 & 1 \end{bmatrix}$.

Because the A in this exercise has three distinct eigenvalues, the only freedom you have in choosing P is to rearrange its columns (and change D accordingly) and to multiply any column of P by -1 (because the eigenvectors must have *unit* length). Notice how the matrix P above differs from the P in the text's answer section.

Warning: The matrices in Exercises 21—24 have only two distinct eigenvalues, so one eigenspace will have to be two-dimensional. (Why?) Be careful when you construct P. Many different choices of P will work.

25. By Theorem 2 (or Theorem 3), A is symmetric. Since A is invertible, a property of transposes shows that $(A^{-1})^T = (A^T)^{-1} = A^{-1}$, so A^{-1} is symmetric. By Theorem 2, again, A^{-1} is orthogonally diagonalizable.

A second argument: By hypothesis, $A = PDP^{-1}$, where P is orthogonal and D is diagonal. Since A is invertible, 0 is not an eigenvalue and D is invertible. Then

$$A^{-1} = (PDP^{-1})^{-1} = (P^{-1})^{-1}D^{-1}P^{-1} = PD^{-1}P^{-1}$$

This shows that A^{-1} is orthogonally diagonalizable.

27. *Note*: To verify the answer given in the text: $A = 8\mathbf{v}_1\mathbf{v}_1^T + 6\mathbf{v}_2\mathbf{v}_2^T + 3\mathbf{v}_3\mathbf{v}_3^T$,

compute $\mathbf{v}_1\mathbf{v}_1^T = \begin{bmatrix} 1/2 & -1/2 & 0 \\ -1/2 & 1/2 & 0 \\ 0 & 0 & 0 \end{bmatrix}$, $\mathbf{v}_2\mathbf{v}_2^T = \begin{bmatrix} 1/6 & 1/6 & -2/6 \\ 1/6 & 1/6 & -2/6 \\ -2/6 & -2/6 & 4/6 \end{bmatrix}$, and

$\mathbf{v}_3\mathbf{v}_3^T = \begin{bmatrix} 1/3 & 1/3 & 1/3 \\ 1/3 & 1/3 & 1/3 \\ 1/3 & 1/3 & 1/3 \end{bmatrix}$. Then $8\mathbf{v}_1\mathbf{v}_1^T + 6\mathbf{v}_2\mathbf{v}_2^T + 3\mathbf{v}_3\mathbf{v}_3^T$ equals

$$\begin{bmatrix} 4 & -4 & 0 \\ -4 & 4 & 0 \\ 0 & 0 & 0 \end{bmatrix} + \begin{bmatrix} 1 & 1 & -2 \\ 1 & 1 & -2 \\ -2 & -2 & 4 \end{bmatrix} + \begin{bmatrix} 1 & 1 & 1 \\ 1 & 1 & 1 \\ 1 & 1 & 1 \end{bmatrix} = \begin{bmatrix} 6 & -2 & -1 \\ -2 & 6 & -1 \\ -1 & -1 & 5 \end{bmatrix} = A$$

33. a. The matrix $B = \mathbf{v}\mathbf{v}^T$ is an outer product, or rank 1 matrix. Given $\mathbf{x}$ in $\mathbb{R}^n$, $B\mathbf{x} = (\mathbf{v}\mathbf{v}^T)\mathbf{x} = \mathbf{v}(\mathbf{v}^T\mathbf{x}) = (\mathbf{v}^T\mathbf{x})\mathbf{v}$, because $\mathbf{v}^T\mathbf{x}$ is a scalar. Using dot products, $B\mathbf{x} = (\mathbf{x}\cdot\mathbf{v})\mathbf{v}$. Since $\mathbf{v}$ is a unit vector, this *is* the orthogonal projection of $\mathbf{x}$ onto $\mathbf{v}$. See Section 7.2.

b. B is symmetric, because $B^T = (\mathbf{v}\mathbf{v}^T)^T = \mathbf{v}^{TT}\mathbf{v}^T = \mathbf{v}\mathbf{v}^T = B$. Also, $B^2 = (\mathbf{v}\mathbf{v}^T)(\mathbf{v}\mathbf{v}^T) = \mathbf{v}(\mathbf{v}^T\mathbf{v})\mathbf{v}^T = \mathbf{v}\mathbf{v}^T = B$, because $\mathbf{v}^T\mathbf{v} = 1$.

MATLAB To check whether a square matrix P is orthogonal, compute **P'*P**. This is faster than checking the individual columns of P. If $P^TP = I$, then P is invertible (by the IMT) and $P^{-1} = P^T$, because P is square.

Don't forget to use **nulbasis(M)** for $M = A - \lambda I$ to speed up eigenvector calculations. If you encounter a two-dimensional eigenspace, with a basis $\{\mathbf{u}_1, \mathbf{u}_2\}$, use the command

```
u2 = u2 - (u2'*u1)/(u1'*u1)*u1
```

or

```
u2 = u2 - proj(u2,u1)
```

to make the new eigenvector $\mathbf{u}_2$ orthogonal to $\mathbf{u}_1$. (Review Section 7.2.) The command **proj** was introduced in the MATLAB note for Section 7.4.

8.2 QUADRATIC FORMS

This section, together with Section 8.1, forms the foundation for the rest of the chapter.

KEY IDEAS

The main point here is to learn how a change of variable, $\mathbf{x} = P\mathbf{u}$, with P an orthogonal matrix, can transform a quadratic form into a new quadratic form with no cross-product terms.

If you study the various classes of quadratic forms (or, equivalently, classes of symmetric matrices), you should learn both the definitions and the characterizations (in Theorem 5) of these classes. Exercise 18 describes another useful way to characterize quadratic forms. (The 2×2 case can be generalized to $n \times n$ matrices.)

SOLUTIONS TO EXERCISES

1. a. $\mathbf{x}^T A\mathbf{x} = [x_1 \quad x_2]\begin{bmatrix} 5 & 1/3 \\ 1/3 & 1 \end{bmatrix}\begin{bmatrix} x_1 \\ x_2 \end{bmatrix} = [x_1 \quad x_2]\begin{bmatrix} 5x_1 + (1/3)x_2 \\ (1/3)x_1 + x_2 \end{bmatrix}$

$= 5x_1^2 + (2/3)x_1x_2 + x_2^2$

b. When $\mathbf{x} = (6,1)$, $\mathbf{x}^T A\mathbf{x} = 5(6)^2 + (2/3)(6)(1) + (1)^2 = 185$

c. When $\mathbf{x} = (1,3)$, $\mathbf{x}^T A\mathbf{x} = 5(1)^2 + (2/3)(1)(3) + (3)^2 = 16$

7. The matrix of the quadratic form is $A = \begin{bmatrix} 1 & 5 \\ 5 & 1 \end{bmatrix}$. The characteristic polynomial is $\lambda^2 - 2\lambda - 24 = (\lambda - 6)(\lambda + 4)$; eigenvalues are 6 and -4.

For $\lambda = 6$: an eigenvector is $\begin{bmatrix} 1 \\ 1 \end{bmatrix}$, normalized: $\mathbf{u}_1 = \frac{1}{\sqrt{2}}\begin{bmatrix} 1 \\ 1 \end{bmatrix}$

For $\lambda = -4$: an eigenvector is $\begin{bmatrix} -1 \\ 1 \end{bmatrix}$, normalized: $\mathbf{u}_2 = \frac{1}{\sqrt{2}}\begin{bmatrix} -1 \\ 1 \end{bmatrix}$

Thus $A = PDP^{-1}$ and $D = P^{-1}AP = P^TAP$ when $P = \frac{1}{\sqrt{2}}\begin{bmatrix} 1 & -1 \\ 1 & 1 \end{bmatrix}$ and $D = \begin{bmatrix} 6 & 0 \\ 0 & -4 \end{bmatrix}$.

The desired change of variable is $\mathbf{x} = P\mathbf{u}$, so that

$$\mathbf{x}^T A\mathbf{x} = (P\mathbf{u})^T A(P\mathbf{u}) = \mathbf{u}^T P^T AP\mathbf{u} = \mathbf{u}^T D\mathbf{u} \qquad (*)$$

$$= 6u_1^2 - 4u_2^2$$

Study Tip: To make the "change of variable" requested in Exercise 7, you should: (1) write the equation $\mathbf{x} = P\mathbf{u}$ and specify P; (2) show the matrix algebra in (*) that produces the new quadratic form; and (3) include the new quadratic form. Find out how much of this information you should supply if a problem like Exercise 7 appears on an exam.

13. The matrix of the quadratic form is $A = \begin{bmatrix} 1 & -3 \\ -3 & 9 \end{bmatrix}$. The characteristic polynomial is $\lambda^2 - 10\lambda = \lambda(\lambda - 10)$; the eigenvalues are 10 and 0. Thus the quadratic form is positive semidefinite. To find the change of variable, proceed as in Exercise 7:

For $\lambda = 10$: an eigenvector is $\begin{bmatrix} 1 \\ -3 \end{bmatrix}$, normalized: $\mathbf{u}_1 = \frac{1}{\sqrt{10}}\begin{bmatrix} 1 \\ -3 \end{bmatrix}$

For $\lambda = 0$: an eigenvector is $\begin{bmatrix} 3 \\ 1 \end{bmatrix}$, normalized: $\mathbf{u}_2 = \frac{1}{\sqrt{10}}\begin{bmatrix} 3 \\ 1 \end{bmatrix}$

Take $P = \frac{1}{\sqrt{10}}\begin{bmatrix} 1 & 3 \\ -3 & 1 \end{bmatrix}$ and $D = \begin{bmatrix} 10 & 0 \\ 0 & 0 \end{bmatrix}$. Since P orthogonally diagonalizes A, the desired change of variable is $\mathbf{x} = P\mathbf{u}$, and

$$\mathbf{x}^T A\mathbf{x} = (P\mathbf{u})^T A(P\mathbf{u}) = \mathbf{u}^T P^T AP\mathbf{u} = \mathbf{u}^T D\mathbf{u} = 10u_1^2$$

The new quadratic form is $10u_1^2$.

19. a. The solution in the text showed that B is symmetric and

$$\mathbf{x}^T A\mathbf{x} = \mathbf{x}^T B^T B\mathbf{x} = (B\mathbf{x})^T B\mathbf{x} = \|B\mathbf{x}\|^2 \geq 0$$

b. To show that A is positive definite, suppose that $\mathbf{x}^T A\mathbf{x} = 0$. Then $\|B\mathbf{x}\|^2 = 0$, and hence $B\mathbf{x} = \mathbf{0}$. If B is invertible, then $\mathbf{x} = \mathbf{0}$, which shows that $\mathbf{x}^T A\mathbf{x} = 0$ only if $\mathbf{x} = 0$. So $\mathbf{x}^T A\mathbf{x}$ is positive definite.

21. The quadratic forms $\mathbf{x}^T A\mathbf{x}$ and $\mathbf{x}^T B\mathbf{x}$ are both positive definite, by Theorem 5, because A and B both have positive eigenvalues. Then for any nonzero $\mathbf{x}$, $\mathbf{x}^T(A + B)\mathbf{x} = \mathbf{x}^T(A\mathbf{x} + B\mathbf{x}) = \mathbf{x}^T A\mathbf{x} + \mathbf{x}^T B\mathbf{x} > 0$, so the quadratic form $\mathbf{x}^T(A + B)\mathbf{x}$ is positive definite. Also, the matrix $A + B$ is symmetric, because $(A + B)^T = A^T + B^T = A + B$. By Theorem 5, $A + B$ has positive eigenvalues.

8.3 CONSTRAINED OPTIMIZATION

This section is important in its own right, since constrained optimization problems arise in many mathematical problems and applications. The main results of the section are also used in the following two sections.

KEY IDEAS

Theorem 6 gives the main idea. The maximum value of a quadratic form $\mathbf{x}^T A\mathbf{x}$ over the set of all unit vectors can be computed by finding the maximum eigenvalue of A; this maximum value is attained at a corresponding eigenvector. Theorem 7 shows the way to generalize this result and characterize the second largest eigenvalue. Example 6 presents a topic that is widely discussed in elementary economics texts.

Checkpoint: Show that if P is an orthogonal matrix and if $\mathbf{x} = P\mathbf{y}$, then $\mathbf{x}$ and $\mathbf{y}$ have the same norm. Thus, $\mathbf{x}$ is a unit vector in $\mathbf{x}^T A\mathbf{x}$ whenever $\mathbf{y}$ is a unit vector in $\mathbf{y}^T D\mathbf{y}$.

SOLUTIONS TO EXERCISES

1. We are given an equality of two quadratic forms:

$$5x_1^2 + 6x_2^2 + 7x_3^2 + 4x_1x_2 - 4x_2x_3 = 9y_1^2 + 6y_2^2 + 3y_3^2$$

The matrix of the left quadratic form obviously is $A = \begin{bmatrix} 5 & 2 & 0 \\ 2 & 6 & -2 \\ 0 & -2 & 7 \end{bmatrix}$.

The equality between the two quadratic forms indicates that the eigenvalues of A are 9,6,3. (Proof: The diagonal matrix D of the quadratic form $9y_1^2 + 6y_2^2 + 3y_3^2$ obviously has eigenvalues 9,6,3. Since A is similar to D, A has the same eigenvalues as D.) The standard calculations produce a unit eigenvector for each eigenvalue. Don't forget to normalize each eigenvector.

$$\lambda = 9:\quad \mathbf{v}_1 = \begin{bmatrix} 1/3 \\ 2/3 \\ -2/3 \end{bmatrix}; \qquad \lambda = 6:\ \mathbf{v}_2 = \begin{bmatrix} 2/3 \\ 1/3 \\ 2/3 \end{bmatrix}; \qquad \lambda = 3:\ \mathbf{v}_3 = \begin{bmatrix} -2/3 \\ 2/3 \\ 1/3 \end{bmatrix}$$

These eigenvectors are mutually orthgonal because they correspond to distinct eigenvalues. So the desired change of variable is

$$\mathbf{x} = P\mathbf{y}, \text{ where } P = \begin{bmatrix} 1/3 & 2/3 & -2/3 \\ 2/3 & 1/3 & 2/3 \\ -2/3 & 2/3 & 1/3 \end{bmatrix}$$

Study Tip: Review the matrix algebra that leads from $\mathbf{x}^T A\mathbf{x}$ to $\mathbf{y}^T D\mathbf{y}$.

7. The matrix of $Q(\mathbf{x}) = -2x_1^2 - x_2^2 + 4x_1x_2 + 4x_2x_3$ is $A = \begin{bmatrix} -2 & 2 & 0 \\ 2 & -1 & 2 \\ 0 & 2 & 0 \end{bmatrix}$.
The hint in the exercise lists 2, -1, and -4 as the eigenvalues. The maximum eigenvalue is 2, not -4, because "maximum" here refers to the eigenvalue that is farthest to the right on the real line. The maximum value of $Q(\mathbf{x})$ (for $\mathbf{x}$ a unit vector) is attained at a unit eigenvector for $\lambda = 2$. Standard calculations produce the eigenvector:

$$\mathbf{u}_1 = \begin{bmatrix} 1/2 \\ 1 \\ 1 \end{bmatrix}, \text{ scaled to } \begin{bmatrix} 1 \\ 2 \\ 2 \end{bmatrix}, \text{ and normalized to } \mathbf{v}_1 = \begin{bmatrix} 1/3 \\ 2/3 \\ 2/3 \end{bmatrix}.$$

Warning: Exercise 7 illustrates the potential error of selecting -4 instead of 2 as the maximum eigenvalue.

12. This exercise can be done by using a theorem, but try to do it by direct computation, using the hint in the text.

13. We may suppose that $m < M$, because the case $m = M$ is trivial. Let t be any number between m and M. We must show that $t = (1 - \alpha)M + \alpha m$ for some number α between 0 and 1. Simple algebra shows that t and α satisfy this equation if and only if $\alpha = (M - t)/(M - m)$, so we can define α by this quotient. Next, we must show that the α so defined lies between 0 and 1. If $t \ge m$, then $M - t \le M - m$, and the quotient that defines α shows that $\alpha \le 1$. Similarly, if $t \le M$, then $M - t \ge 0$ and so $\alpha \ge 0$. Thus α ranges between 0 and 1.
Next, let $\mathbf{x} = \sqrt{1-\alpha}\,\mathbf{v}_1 + \sqrt{\alpha}\,\mathbf{v}_n$. Observe that the vectors $\sqrt{1-\alpha}\,\mathbf{v}_1$ and $\sqrt{\alpha}\,\mathbf{v}_n$ are orthogonal because they are eigenvectors for different eigenvalues. By the Pythagorean Theorem,

$$\mathbf{x}^T\mathbf{x} = \|\mathbf{x}\|^2 = \|\sqrt{1-\alpha}\,\mathbf{v}_1\|^2 + \|\sqrt{\alpha}\,\mathbf{v}_n\|^2$$
$$= |1 - \alpha|\,\|\mathbf{v}_1\|^2 + |\alpha|\,\|\mathbf{v}_n\|^2 = (1 - \alpha) + \alpha = 1$$

because $\mathbf{v}_1$ and $\mathbf{v}_n$ are unit vectors and $0 \le \alpha \le 1$. Also, using the facts that $A\mathbf{v}_1 = M\mathbf{v}_1$, $A\mathbf{v}_n = m\mathbf{v}_n$, and $\mathbf{v}_1$ and $\mathbf{v}_n$ are orthogonal, we have

$$\mathbf{x}^TA\mathbf{x} = [\sqrt{1-\alpha}\,\mathbf{v}_1 + \sqrt{\alpha}\,\mathbf{v}_n]^T A[\sqrt{1-\alpha}\,\mathbf{v}_1 + \sqrt{\alpha}\,\mathbf{v}_n]$$
$$= [\sqrt{1-\alpha}\,\mathbf{v}_1 + \sqrt{\alpha}\,\mathbf{v}_n]^T[M\sqrt{1-\alpha}\,\mathbf{v}_1 + m\sqrt{\alpha}\,\mathbf{v}_n]$$
$$= |1 - \alpha|M\mathbf{v}_1^T\mathbf{v}_1 + |\alpha|m\mathbf{v}_n^T\mathbf{v}_n = (1 - \alpha)M + \alpha m = t$$

This shows that the quadratic form $\mathbf{x}^T A\mathbf{x}$ assumes every value between m and M as $\mathbf{x}$ varies over all unit vectors.

Answer to Checkpoint: Show that $\|\mathbf{x}\|^2 = \|\mathbf{y}\|^2$. Use the basic relation between the norm and the inner product, and use the fact that $P^T P = I$ because P is an orthogonal matrix: $\|\mathbf{x}\|^2 = \|P\mathbf{y}\|^2 = (P\mathbf{y})^T P\mathbf{y} = \mathbf{y}^T P^T P\mathbf{y} = \mathbf{y}^T\mathbf{y} = \|\mathbf{y}\|^2$.

8.4 THE SINGULAR VALUE DECOMPOSITION

This section is the capstone of the text. It completes the story of the linear transformation $\mathbf{x} \mapsto A\mathbf{x}$ for a general $m \times n$ matrix A, and in so doing gives you an opportunity to review many basic concepts from Chapters 5-8. In addition, this section opens the door into the modern world of applied linear algebra. An understanding of the singular value decomposition is essential for advanced work in science and engineering that requires matrix computations.

KEY IDEAS

The first singular value σ_1 of an $m \times n$ matrix A is the maximum of $\|A\mathbf{x}\|$ over all unit vectors. This maximum value is attained at a unit eigenvector $\mathbf{v}_1$ of $A^T A$ corresponding to the largest eigenvalue λ_1 of $A^T A$. The other singular values could be characterized similarly. The following algorithm produces the singular value decomposition for A. (As mentioned in the text, other more reliable methods are used in professional software.)

Procedure for Computing a Singular Value Decomposition

1. Find an orthonormal basis $\{\mathbf{v}_1, \ldots, \mathbf{v}_n\}$ for $\mathbb{R}^n$ consisting of eigenvectors of $A^T A$, arranged so that the associated eigenvalues satisfy $\lambda_1 \geq \cdots \geq \lambda_r > 0$ and $\lambda_{r+1} = \cdots = 0$, where $r = \text{rank}\, A$.
2. Construct the $n \times n$ orthogonal matrix $V = [\mathbf{v}_1 \ \cdots \ \mathbf{v}_n]$.
3. Let $\sigma_j = \sqrt{\lambda_j}$ $(1 \leq j \leq r)$, and construct the $m \times n$ diagonal matrix Σ whose (j, j)-entry is σ_j $(1 \leq j \leq r)$ and has zeros elsewhere.
4. The set $\{A\mathbf{v}_1, \ldots, A\mathbf{v}_r\}$ is orthogonal and $\sigma_j = \|A\mathbf{v}_j\|$. Compute $\mathbf{u}_j = (1/\sigma_j)A\mathbf{v}_j$ $(1 \leq j \leq r)$.

5. Extend $\{\mathbf{u}_1, \ldots, \mathbf{u}_r\}$ to an orthonormal basis $\{\mathbf{u}_1, \ldots, \mathbf{u}_m\}$ for $\mathbb{R}^m$. Construct the $m \times m$ orthogonal matrix $U = [\mathbf{u}_1 \quad \cdots \quad \mathbf{u}_m]$.
6. $A = U\Sigma V^T$.

The diagram below shows how the vectors $\mathbf{v}_j$ and $\mathbf{u}_j$ are affected by A and A^T. Since $A\mathbf{v}_j = \sigma_j\mathbf{u}_j$ (by construction of $\mathbf{u}_j$), we have $A^TA\mathbf{v}_j = \sigma_j A^T\mathbf{u}_j$. But $(A^TA)\mathbf{v}_j = \sigma_j^2\mathbf{v}_j$ (because $\mathbf{v}_j$ is an eigenvector of A^TA), and so $A^T\mathbf{u}_j = \sigma_j\mathbf{v}_j$.

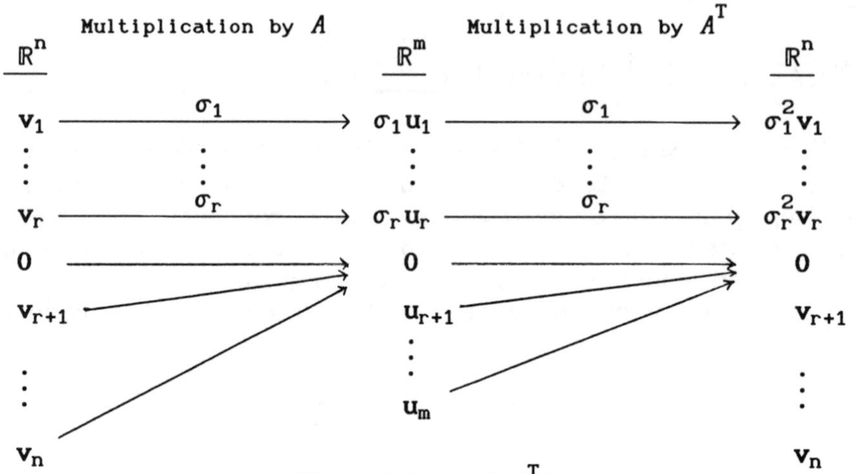

The action of A^TA

The figure below will help you remember the details of Example 6, which describes orthonormal bases for the four fundamental subspaces for A.

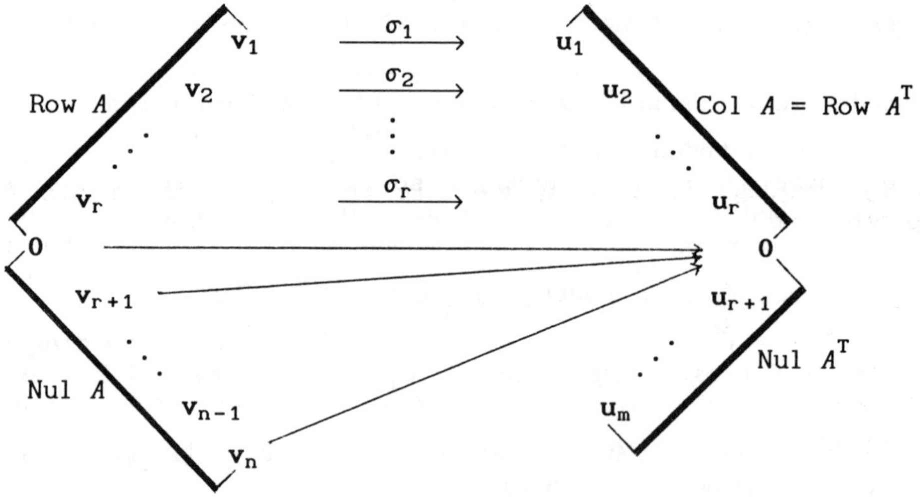

The Four Fundamental Subspaces and the action of A

If you like this diagram, you can modify it to show the action of the pseudoinverse A^+, discussed in Example 7. Simply change the arrows appropriately and replace each σ_j by $1/\sigma_j$.

SOLUTIONS TO EXERCISES

1. $A = \begin{bmatrix} 1 & 0 \\ 0 & -3 \end{bmatrix}$, $A^TA = \begin{bmatrix} 1 & 0 \\ 0 & 9 \end{bmatrix}$. The eigenvalues and eigenvectors are obvious:

$$\lambda_1 = 9: \quad \mathbf{v}_1 = \begin{bmatrix} 0 \\ 1 \end{bmatrix}; \qquad \lambda_2 = 1: \quad \mathbf{v}_2 = \begin{bmatrix} 1 \\ 0 \end{bmatrix}$$

(Remember to arrange the eigenvalues in decreasing order.) Thus

$$V = \begin{bmatrix} 0 & 1 \\ 1 & 0 \end{bmatrix}$$

The singular values are $\sigma_1 = \sqrt{9} = 3$ and $\sigma_2 = 1$. The matrix Σ is the same shape as A and

$$\Sigma = \begin{bmatrix} \sigma_1 & 0 \\ 0 & \sigma_2 \end{bmatrix} = \begin{bmatrix} 3 & 0 \\ 0 & 1 \end{bmatrix}$$

Next, compute

$$A\mathbf{v}_1 = \begin{bmatrix} 1 & 0 \\ 0 & -3 \end{bmatrix}\begin{bmatrix} 0 \\ 1 \end{bmatrix} = \begin{bmatrix} 0 \\ -3 \end{bmatrix}, \quad A\mathbf{v}_2 = \begin{bmatrix} 1 & 0 \\ 0 & -3 \end{bmatrix}\begin{bmatrix} 1 \\ 0 \end{bmatrix} = \begin{bmatrix} 1 \\ 0 \end{bmatrix}$$

and normalize:

$$\mathbf{u}_1 = \frac{1}{\sigma_1}A\mathbf{v}_1 = \frac{1}{3}\begin{bmatrix} 0 \\ -3 \end{bmatrix} = \begin{bmatrix} 0 \\ -1 \end{bmatrix}, \quad \mathbf{u}_2 = \frac{1}{\sigma_2}\begin{bmatrix} 1 \\ 0 \end{bmatrix} = \begin{bmatrix} 1 \\ 0 \end{bmatrix}$$

Finally, $\{\mathbf{u}_1, \mathbf{u}_2\}$ is already a basis for $\mathbb{R}^2$, so the basis for $\mathbb{R}^2$ is complete, and

$$U = \begin{bmatrix} 0 & 1 \\ -1 & 0 \end{bmatrix}, \text{ and } A = U\Sigma V^T = \begin{bmatrix} 0 & 1 \\ -1 & 0 \end{bmatrix}\begin{bmatrix} 3 & 0 \\ 0 & 1 \end{bmatrix}\begin{bmatrix} 0 & 1 \\ 1 & 0 \end{bmatrix}$$

This happens to equal V.

7. $A = \begin{bmatrix} 2 & -1 \\ 2 & 2 \end{bmatrix}$. $A^TA = \begin{bmatrix} 8 & 2 \\ 2 & 5 \end{bmatrix}$. Find the eigenvalues of A^TA from the characteristic equation.

$$0 = \lambda^2 - 13\lambda + 36 = (\lambda - 9)(\lambda - 4); \quad \lambda_1 = 9, \ \lambda_2 = 4$$

Corresponding unit eigenvectors for A^TA (calculations omitted) are:

$$\lambda_1 = 9: \quad \mathbf{v}_1 = \begin{bmatrix} 2/\sqrt{5} \\ 1/\sqrt{5} \end{bmatrix}; \qquad \lambda_2 = 4: \quad \mathbf{v}_2 = \begin{bmatrix} -1/\sqrt{5} \\ 2/\sqrt{5} \end{bmatrix}$$

Take

$$V = \begin{bmatrix} 2/\sqrt{5} & -1/\sqrt{5} \\ 1/\sqrt{5} & 2/\sqrt{5} \end{bmatrix}$$

The singular values are $\sigma_1 = \sqrt{9} = 3$ and $\sigma_2 = \sqrt{4} = 2$. The matrix Σ is the same shape as A and $\Sigma = \begin{bmatrix} \sigma_1 & 0 \\ 0 & \sigma_2 \end{bmatrix} = \begin{bmatrix} 3 & 0 \\ 0 & 2 \end{bmatrix}$. Next, compute

$$A\mathbf{v}_1 = \begin{bmatrix} 2 & -1 \\ 2 & 2 \end{bmatrix}\begin{bmatrix} 2/\sqrt{5} \\ 1/\sqrt{5} \end{bmatrix} = \begin{bmatrix} 3/\sqrt{5} \\ 6/\sqrt{5} \end{bmatrix}, \qquad A\mathbf{v}_2 = \begin{bmatrix} 2 & -1 \\ 2 & 2 \end{bmatrix}\begin{bmatrix} -1/\sqrt{5} \\ 2/\sqrt{5} \end{bmatrix} = \begin{bmatrix} -4/\sqrt{5} \\ 2/\sqrt{5} \end{bmatrix}$$

To check your work at this point, verify that $A\mathbf{v}_1$ and $A\mathbf{v}_2$ are orthogonal. (They are.) Then normalize:

$$\mathbf{u}_1 = \frac{1}{\sigma_1}A\mathbf{v}_1 = \frac{1}{3}\begin{bmatrix} 3/\sqrt{5} \\ 6/\sqrt{5} \end{bmatrix} = \begin{bmatrix} 1/\sqrt{5} \\ 2/\sqrt{5} \end{bmatrix}, \qquad \mathbf{u}_2 = \frac{1}{\sigma_2} = \frac{1}{2}\begin{bmatrix} -4/\sqrt{5} \\ 2/\sqrt{5} \end{bmatrix} = \begin{bmatrix} -2/\sqrt{5} \\ 1/\sqrt{5} \end{bmatrix}$$

Since $\{\mathbf{u}_1, \mathbf{u}_2\}$ is a basis for $\mathbb{R}^2$, take

$$U = \begin{bmatrix} 1/\sqrt{5} & -2/\sqrt{5} \\ 2/\sqrt{5} & 1/\sqrt{5} \end{bmatrix}$$

Thus $A = U\Sigma V^T = \begin{bmatrix} 1/\sqrt{5} & -2/\sqrt{5} \\ 2/\sqrt{5} & 1/\sqrt{5} \end{bmatrix}\begin{bmatrix} 3 & 0 \\ 0 & 2 \end{bmatrix}\begin{bmatrix} 2/\sqrt{5} & 1/\sqrt{5} \\ -1/\sqrt{5} & 2/\sqrt{5} \end{bmatrix}$. (Use V^T, not V.)

Study Tip: Your answer for a singular value decomposition may differ from that given in the text. To check your work, compute AV and $U\Sigma$. If $AV = U\Sigma$, then $A = U\Sigma V^T$ and your answer is correct (provided V truly is an orthogonal matrix).

13. The matrix A^TA is 3×3. Because the text has not given you practice computing and solving a cubic characteristic equation, the *Hint* suggests that you consider A^T instead of A. (You are free to work on A itself, if you prefer.) Using A^T, we compute

$$(A^T)^TA^T = AA^T = \begin{bmatrix} 3 & 2 & 2 \\ 2 & 3 & -2 \end{bmatrix}\begin{bmatrix} 3 & 2 \\ 2 & 3 \\ 2 & -2 \end{bmatrix} = \begin{bmatrix} 17 & 8 \\ 8 & 17 \end{bmatrix}$$

The characteristic equation is

$$0 = \lambda^2 - 34\lambda + 225 = (\lambda - 25)(\lambda - 9); \quad \lambda_1 = 25, \ \lambda_2 = 9$$

The corresponding unit eigenvectors and the matrix V are

$$\lambda_1 = 25\colon\ \mathbf{v}_1 = \begin{bmatrix} 1/\sqrt{2} \\ 1/\sqrt{2} \end{bmatrix}; \quad \lambda_2 = 9\colon\ \mathbf{v}_2 = \begin{bmatrix} -1/\sqrt{2} \\ 1/\sqrt{2} \end{bmatrix}; \quad V = \begin{bmatrix} 1/\sqrt{2} & -1/\sqrt{2} \\ 1/\sqrt{2} & 1/\sqrt{2} \end{bmatrix}$$

The singular values are $\sigma_1 = 5$ and $\sigma_2 = 3$. Thus Σ is $\begin{bmatrix} 3 & 0 \\ 0 & 2 \\ 0 & 0 \end{bmatrix}$, the same size as A^T. To get $\mathbf{u}_1$ and $\mathbf{u}_2$, compute $A^T\mathbf{v}_1$ and $A^T\mathbf{v}_2$,

$$A^T[\mathbf{v}_1 \ \ \mathbf{v}_2] = \begin{bmatrix} 3 & 2 \\ 2 & 3 \\ 2 & -2 \end{bmatrix} \begin{bmatrix} 1/\sqrt{2} & -1/\sqrt{2} \\ 1/\sqrt{2} & 1/\sqrt{2} \end{bmatrix} = \begin{bmatrix} 5/\sqrt{2} & -1/\sqrt{2} \\ 5/\sqrt{2} & 1/\sqrt{2} \\ 0 & -4/\sqrt{2} \end{bmatrix}$$

and normalize:

$$\mathbf{u}_1 = \begin{bmatrix} 1/\sqrt{2} \\ 1/\sqrt{2} \\ 0 \end{bmatrix}, \quad \mathbf{u}_2 = \begin{bmatrix} -1/\sqrt{18} \\ 1/\sqrt{18} \\ -4/\sqrt{18} \end{bmatrix}$$

We need one more vector, orthogonal to $\mathbf{u}_1$ and $\mathbf{u}_2$. So write the equations $\mathbf{u}_1^T\mathbf{x} = 0$ and $\mathbf{u}_2^T\mathbf{x} = 0$ and solve for $\mathbf{x}$. Simpler equations are

$$\begin{aligned} \sqrt{2}\,\mathbf{u}_1^T\mathbf{x} &= 0 \\ \sqrt{18}\,\mathbf{u}_2^T\mathbf{x} &= 0 \end{aligned} \qquad \text{or} \qquad \begin{aligned} x_1 + x_2 \phantom{{}- 4x_3} &= 0 \\ -x_1 + x_2 - 4x_3 &= 0 \end{aligned}$$

The solution is $x_1 = -2x_3$, $x_2 = 2x_3$, x_3 free. A suitable unit vector is

$$\mathbf{u}_3 = \begin{bmatrix} -2/3 \\ 2/3 \\ 1/3 \end{bmatrix}$$

Thus an SVD of A^T is

$$A^T = [\mathbf{u}_1 \ \ \mathbf{u}_2 \ \ \mathbf{u}_3] \begin{bmatrix} 5 & 0 \\ 0 & 3 \\ 0 & 0 \end{bmatrix} [\mathbf{v}_1 \ \ \mathbf{v}_2]^T$$

So an SVD of A appears by taking transposes:

$$A = \begin{bmatrix} 1/\sqrt{2} & -1/\sqrt{2} \\ 1/\sqrt{2} & 1/\sqrt{2} \end{bmatrix} \begin{bmatrix} 5 & 0 & 0 \\ 0 & 3 & 0 \end{bmatrix} \begin{bmatrix} 1/\sqrt{2} & 1/\sqrt{2} & 0 \\ -1/\sqrt{18} & 1/\sqrt{18} & -4/\sqrt{18} \\ -2/3 & 2/3 & 1/3 \end{bmatrix}$$

This *is* an SVD because the outside matrices are orthogonal matrices, and the center matrix is a diagonal matrix of the proper type. Another way to find $\mathbf{u}_3$ is to realize that $\mathbf{u}_1$ and $\mathbf{u}_2$ form an orthonormal basis for Col A^T = Row A. The remaining $\mathbf{u}_3$ must be a basis for $(\text{Row } A)^\perp$ = Nul A.

15. Let $A = U\Sigma V^T$. Then

$$A^TA = (U\Sigma V^T)^TU\Sigma V^T = V\Sigma^TU^TU\Sigma V^T$$

$$= V(\Sigma^T\Sigma)V^{-1} \quad \text{Because U and V are orthogonal}$$

If $\sigma_1,\ldots,\sigma_r$ are the nonzero diagonal entries in Σ, then $\Sigma^T\Sigma$ is diagonal, with diagonal entries $\sigma_1^2,\ldots,\sigma_r^2$, and possibly some zeros. Thus V diagonalizes A^TA. By the Diagonalization Theorem in Section 6.3, the columns of V are eigenvectors of A^TA, and $\sigma_1^2,\ldots,\sigma_r^2$ are the nonzero eigenvalues of A^TA. Hence $\sigma_1,\ldots,\sigma_r$ are the nonzero singular values of A. A similar calculation of AA^T shows that the columns of U are eigenvectors of AA^T.

19. From the proof of Theorem 10, $U\Sigma = [\sigma_1\mathbf{u}_1 \ \cdots \ \sigma_r\mathbf{u}_r \ \ \mathbf{0} \ \cdots \ \mathbf{0}]$. The column-row expansion of a matrix product shows that

$$A = (U\Sigma)V^T = (U\Sigma)\begin{bmatrix}\mathbf{v}_1^T \\ \vdots \\ \mathbf{v}_n^T\end{bmatrix} = \sigma_1\mathbf{u}_1\mathbf{v}_1^T + \cdots + \sigma_r\mathbf{u}_r\mathbf{v}_r^T$$

This expansion generalizes the spectral decomposition in Section 8.1.

Study Tip: In Exercise 19, the *left* singular vectors are the columns $\mathbf{u}_1,\ldots,\mathbf{u}_m$ of the *left* factor U in $U\Sigma V^T$, but the *right* singular vectors are the columns of V, *not* V^T.

21. The right singular vector $\mathbf{v}_1$ is an eigenvector for the largest eigenvalue λ_1 of A^TA. By Theorem 7 in Section 8.3, the second largest eigenvalue, λ_2, is the maximum of $\mathbf{x}^T(A^TA)\mathbf{x}$ over all unit vectors orthogonal to $\mathbf{v}_1$. Since $\mathbf{x}^T(A^TA)\mathbf{x} = \|A\mathbf{x}\|^2$, the square root of λ_2, which is the second singular value of A, is the maximum of $\|A\mathbf{x}\|$ over all unit vectors orthogonal to $\mathbf{v}_1$.

23. Consider the SVD for the standard matrix of T, say, $A = U\Sigma V^T$. Let $\mathcal{B} = \{\mathbf{v}_1,\ldots,\mathbf{v}_n\}$ and $\mathcal{C} = \{\mathbf{u}_1,\ldots,\mathbf{u}_m\}$ be bases constructed from the columns of V and U, respectively. Observe that, since the columns of V are orthonormal, $V^T\mathbf{v}_j = \mathbf{e}_j$, where $\mathbf{e}_j$ is the jth column of the $n\times n$ identity matrix. To find the matrix of T relative to $\mathcal{B}$ and $\mathcal{C}$, compute

$$T(\mathbf{v}_j) = A\mathbf{v}_j = U\Sigma V^T\mathbf{v}_j = U\Sigma\mathbf{e}_j = U\sigma_j\mathbf{e}_j = \sigma_j U\mathbf{e}_j = \sigma_j\mathbf{u}_j$$

So $[T(\mathbf{v}_j)]_{\mathscr{C}} = \sigma_j$. Formula (4) in the discussion at the beginning of Section 6.4 shows that the "diagonal" matrix Σ is the matrix of T relative to $\mathscr{B}$ and $\mathscr{C}$.

MATLAB The command **[P D] = eig(A'*A)** produces an orthogonal matrix P of eigenvectors and a diagonal matrix D of eigenvalues of A^TA, but the eigenvalues in D may not be in decreasing order. In such a case, you will have to rearrange things to form V and Σ (denoted below by S).

For instance, if P is 3×3, the command

```
V = P(:,[1 3 2])
```

interchanges columns 2 and 3 of P to form V. The commands

```
S = zeros(A); S(2,2) = sqrt(D(3,3))
```

produce a zero matrix for "Σ" the same size as A and place the square root of the (3,3)-entry of D into the (2,2)-entry of S. Other diagonal entries for S can be entered similarly. To form U for the SVD, normalize the nonzero columns of $A*V$. If U needs more columns, use the method of Example 4.

8.5 APPLICATIONS TO IMAGE PROCESSING AND STATISTICS

If you find remote sensing or image processing interesting, or if you plan to use multivariate statistics later in your career, then you will want to study this section thoroughly. You may have difficulty finding an elementary explanation of this material elsewhere. The idea for the application to image processing came from a student, a geography major who was taking an undergraduate course in remote sensing while she was in my linear algebra class. The book by Lillesand and Kiefer, referenced in the text, was one of the texts for her course.

KEY IDEAS

The **first principal component** of the data in the matrix of observations is a unit eigenvector $\mathbf{v}_1$ corresponding to the largest eigenvalue of the covariance matrix S. If $\mathbf{v}_1 = (c_1,\dots,c_p)$, then the entries in $\mathbf{v}_1$ are weights in a linear combination of the original variables, $x_1, \dots, x_p$, that creates a new variable u_1 (sometimes called a composite score or *index*):

$$u_1 = \mathbf{v}_1^T\mathbf{X} = c_1x_1 + \cdots + c_px_p$$

The variance of the values of this index is the largest possible among all indices whose coefficients $c_1, \dots, c_p$ form a unit vector. (The variance of u_1 is the largest eigenvalue of S.) The **second principal component** is the unit eigenvector corresponding to the second largest eigenvalue of S. The entries in the second principal component determine the index with greatest variance among all possible indices (from a unit vector) that are uncorrelated (in a statistical sense) with u_1. Additional principal components are defined similarly.

Checkpoints: (1) If the variables x_1 and x_3 are uncorrelated, what can you say about the covariance matrix S? (2) What is the covariance matrix of the new variables $u_1, \dots, u_p$ formed from the principal components of S?

SOLUTIONS TO EXERCISES

1. The matrix of observations is $X = \begin{bmatrix} 19 & 22 & 6 & 3 & 2 & 20 \\ 12 & 6 & 9 & 15 & 13 & 5 \end{bmatrix}$, and the sample mean **M** is $\begin{bmatrix} 12 \\ 10 \end{bmatrix}$. Subtract **M** from each column of X to obtain

$$B = \begin{bmatrix} 7 & 10 & -6 & -9 & -10 & 8 \\ 2 & -4 & -1 & 5 & 3 & -5 \end{bmatrix}$$

The sample covariance matrix is

$$S = \frac{1}{N-1}BB^T = \frac{1}{5}\begin{bmatrix} 7 & 10 & -6 & -9 & -10 & 8 \\ 2 & -4 & -1 & 5 & 3 & -5 \end{bmatrix}\begin{bmatrix} 7 & 2 \\ 10 & -4 \\ -6 & -1 \\ -9 & 5 \\ -10 & 3 \\ 8 & -5 \end{bmatrix}$$

$$= \frac{1}{5}\begin{bmatrix} 430 & -135 \\ -135 & 80 \end{bmatrix} = \begin{bmatrix} 86 & -27 \\ -27 & 16 \end{bmatrix}$$ Usually, S contains decimals.

Study Tip: Note that the formula for the sample mean involves division by N, but for statistical reasons, the covariance matrix formula involves division by $N - 1$.

7. Let x_1, x_2 denote the variables for the two-dimensional data in Exercise 1. The characteristic equation of the covariance matrix S from Exercise

1 is $\lambda^2 - 102\lambda + 647 = 0$. By the quadratic formula, the roots of this equation are $\lambda_1 = 95.20$ and $\lambda_2 = 6.80$ (to two decimal places). The first principal component of the data is a unit eigenvector corresponding to λ_1, which turns out to be (-.95, .32), or (.95, -.32). The two possible choices for the new variable are $u_1 = -.95x_1 + .32x_2$ and $u_1 = .95x_1 - .32x_2$. The variance of u_1 is 95.20, while the total variance is 95.20 + 6.80 = 102. Since 95.20/102 = .933, the new variable u_1 explains about 93.3% of the variance in the data.

11. a. The solution in the text shows that the $\mathbf{U}_j$ are in mean-deviation form, where $\mathbf{U}_j = P^T\mathbf{X}_j$ for some $p \times p$ matrix P.

b. By part (a), the covariance matrix of $\mathbf{U}_1, \ldots, \mathbf{U}_N$ is

$$\frac{1}{N-1}[\mathbf{U}_1 \ \cdots \ \mathbf{U}_N][\mathbf{U}_1 \ \cdots \ \mathbf{U}_N]^T$$

$$= \frac{1}{N-1}P^T[\mathbf{X}_1 \ \cdots \ \mathbf{X}_N]\left(P^T[\mathbf{X}_1 \ \cdots \ \mathbf{X}_N]\right)^T$$

$$= P^T\left(\frac{1}{N-1}[\mathbf{X}_1 \ \cdots \ \mathbf{X}_N][\mathbf{X}_1 \ \cdots \ \mathbf{X}_N]^T\right)P$$

$$= P^TSP$$

because $\mathbf{X}_1, \ldots, \mathbf{X}_N$ are in mean-deviation form.

13. Let $\mathbf{M}$ be the sample mean of the data, and for $k = 1, \ldots, N$, write $\hat{\mathbf{X}}_k$ for $\mathbf{X}_k - \mathbf{M}$. Let $B = [\hat{\mathbf{X}}_1 \ \cdots \ \hat{\mathbf{X}}_N]$, the matrix of observations in mean deviation form. By the column-row expansion of BB^T, the sample covariance matrix is

$$S = \frac{1}{N-1}BB^T = \frac{1}{N-1}[\hat{\mathbf{X}}_1 \ \cdots \ \hat{\mathbf{X}}_N]\begin{bmatrix}\hat{\mathbf{X}}_1^T \\ \vdots \\ \hat{\mathbf{X}}_N^T\end{bmatrix}$$

$$= \frac{1}{N-1}\sum_{1}^{N}\hat{\mathbf{X}}_k\hat{\mathbf{X}}_k^T = \frac{1}{N-1}\sum_{1}^{N}(\mathbf{X}_k - \mathbf{M})(\mathbf{X}_k - \mathbf{M})^T$$

Answers to Checkpoints: (1) The (1,3)- and (3,1)-entries of S are zero. (2) The covariance matrix of $u_1, \ldots, u_p$ is the diagonal matrix formed from the eigenvalues of S. This matrix is diagonal because the new variables are pairwise uncorrelated.

MATLAB The command **mean(X')** produces a row vector whose entries list the averages of the rows of X, and **diag(mean(X'))** creates a diagonal matrix whose diagonal entries are the row averages of X. (Be careful not to use **mean(X)** , which lists the averages of the *columns* of X.) Finally, the command **diag(mean(X'))*ones(X)** creates a matrix the size of X, whose columns are all the same, each one listing the row averages of X. To convert the data in X into mean-deviation form, use

```
B = X - diag(mean(X'))*ones(X)
```

The sample covariance matrix is produced by

```
S = B*B'/(N-1)
```

The principal component data you need is contained in P and D, where

```
[P  D] = eig(S)
```

CHAPTER 8 GLOSSARY CHECKLIST

Check your knowledge by attempting to write definitions of the terms below. Then compare your work with the definitions given in the text's Glossary. Ask your instructor which definitions, if any, might appear on a test.

condition number (of A): The ... [number], where ....

covariance (of variables x_i and x_j, for $i \neq j$): The entry ... in the ... matrix ... for a matrix of observations, where x_i and x_j vary over the ....

covariance matrix (or **sample covariance matrix**): The $p \times p$ matrix S defined by S = ..., where B is a $p \times N$ matrix of observations ....

indefinite matrix: A symmetric matrix A such that ....

indefinite quadratic form: A quadratic form Q such that $Q(\mathbf{x})$ ....

left singular vectors (of A): The columns of ... in the ....

matrix of observations: A $p \times N$ matrix whose columns are ..., each column listing ....

mean-deviation form (of a matrix of observations): A matrix whose ... vectors are ....

Moore-Penrose inverse: *See* pseudoinverse.

negative definite matrix: A symmetric matrix A such that ....

negative definite quadratic form: A quadratic form Q such that $Q(\mathbf{x})$ ....

negative semidefinite matrix: A symmetric matrix A such that ....

negative semidefinite quadratic form: A quadratic form Q such that ....

orthogonally diagonalizable: A matrix A that admits a factorization, A = ..., with ... and ....

positive definite matrix: A symmetric matrix A such that ....

positive definite quadratic form: A quadratic form Q such that $Q(\mathbf{x})$ ....

positive semidefinite matrix: A symmetric matrix A such that ....

positive semidefinite quadratic form: A quadratic form Q such that ....

principal axes (of a quadratic form $\mathbf{x}^T A\mathbf{x}$): The ... columns of an ... matrix P such that ....

principal components (of the data in a matrix of observations B): The ... vectors of a sample covariance matrix S for B, with the ... vectors arranged so that ....

projection matrix (or **orthogonal projection matrix**): A symmetric matrix B such that .... The simplest example is B = ...

pseudoinverse (of A): The matrix ..., when ... is a reduced singular value decomposition of A.

quadratic form: A function Q defined for $\mathbf{x}$ in $\mathbb{R}^n$ by $Q(\mathbf{x})$ = ..., where A is an $n \times n$ ... matrix A (called the ...).

reduced singular value decomposition: A factorization A = ..., for an $m \times n$ matrix A of rank r, where ... is ... with ... columns, ... is ... with ..., and ... is ... with ... columns.

right singular vectors (of A): The columns of ... in the singular value decomposition A = ...

row sum: ....

sample mean: The ... of a set of vectors, $\mathbf{X}_1, \ldots, \mathbf{X}_N$, given by $\mathbf{M}$ = ...

singular value decomposition (of an $m \times n$ matrix A): A = ..., where ... is an ... matrix, ... is an ... matrix, and ... is an ...matrix with ....

singular values (of A): The ... of the eigenvalues of ..., arranged ....

spectral decomposition (of A): A representation A = ... where ....

symmetric matrix: A matrix A such that ....

total variance: The ... of the ... matrix S of a matrix of ....

uncorrelated variables: Any two variables x_i and x_j (with $i \neq j$) that range over the ith and jth coordinates of the observation vectors in an observation matrix, such that ....

variance of a variable x_j: The ... entry ... in the ... matrix S for a matrix of ..., where x_j varies over the ....

INDEX OF MATLAB COMMANDS AND OPERATIONS